Werkstofftechnische Berichte | Reports of Materials Science and Engineering

Reihe herausgegeben von

Frank Walther, Lehrstuhl für Werkstoffprüftechnik (WPT), TU Dortmund, Dortmund, Nordrhein-Westfalen, Deutschland

In den Werkstofftechnischen Berichten werden Ergebnisse aus Forschungspro-jekten veröffentlicht, die am Lehrstuhl für Werkstoffprüftechnik (WPT) der Technischen Universität Dortmund in den Bereichen Materialwissenschaft und Werkstofftechnik sowie Mess- und Prüftechnik bearbeitet wurden. Die Forschungsergebnisse bilden eine zuverlässige Datenbasis für die Konstruktion, Fertigung und Überwachung von Hochleistungsprodukten für unterschiedliche wirtschaftliche Branchen. Die Arbeiten geben Einblick in wissenschaftliche und anwendungsorientierte Fragestellungen, mit dem Ziel, strukturelle Integrität durch Werkstoffverständnis unter Berücksichtigung von Ressourceneffizienz zu gewährleisten.

Optimierte Analyse-, Auswerte- und Inspektionsverfahren werden als Entschei-dungshilfe bei der Werkstoffauswahl und -charakterisierung, Qualitätskontrolle und Bauteilüberwachung sowie Schadensanalyse genutzt. Neben der Werkstof-fqualifizierung und Fertigungsprozessoptimierung gewinnen Maßnahmen des Structural Health Monitorings und der Lebensdauervorhersage an Bedeutung. Bewährte Techniken der Werkstoff- und Bauteilcharakterisierung werden weit-erentwickelt und ergänzt, um den hohen Ansprüchen neuentwickelter Produk-tionsprozesse und Werkstoffsysteme gerecht zu werden.

Reports of Materials Science and Engineering aims at the publication of results of research projects carried out at the Chair of Materials Test Engineering (WPT) at TU Dortmund University in the fields of materials science and engineering as well as measurement and testing technologies. The research results contribute to a reliable database for the design, production and monitoring of high-performance products for different industries. The findings provide an insight to scientific and applied issues, targeted to achieve structural integrity based on materials understanding while considering resource efficiency.

Optimized analysis, evaluation and inspection techniques serve as decision guid-ance for material selection and characterization, quality control and component monitoring, and damage analysis. Apart from material qualification and produc-tion process optimization, activities concerning structural health monitoring and service life prediction are in focus. Established techniques for material and com-ponent characterization are aimed to be improved and completed, to match the high demands of novel production processes and material systems.

Mustafa Mamduh Mustafa Awd

Machine Learning Algorithm for Fatigue Fields in Additive Manufacturing

Springer Vieweg

Mustafa Mamduh Mustafa Awd
Dortmund, Germany

Publication as Doctoral Thesis in Faculty of Mechanical Engineering of TU Dortmund University.
Location of doctorate: Dortmund
Date of oral presentation: 05.09.2022
Chairman: Prof. Dr. Moritz Schulze Darup
1. Reviewer: Prof. Dr.-Ing. habil. Frank Walther
2. Reviewer: Prof. Dr.-Ing. Sebastian Münstermann
Assessor: Prof. Dr.-Ing. Dr.-Ing. E. h. A. Erman Tekkaya

ISSN 2524-4809   ISSN 2524-4817 (electronic)
Werkstofftechnische Berichte | Reports of Materials Science and Engineering
ISBN 978-3-658-40236-5   ISBN 978-3-658-40237-2 (eBook)
https://doi.org/10.1007/978-3-658-40237-2

This Springer Vieweg imprint is published by the registered company Springer Fachmedien Wiesbaden GmbH, part of Springer Nature.
The registered company address is: Abraham-Lincoln-Str. 46, 65189 Wiesbaden, Germany

Foreword

The impact of process-specific characteristics such as surface roughness, remnant porosity, residual stresses, and microstructure on process-oriented characteristics of engineering materials has an effect on the property profile of emerging structures. TU Dortmund University's Chair of Materials Test Engineering (WPT) capitalizes on the promise of additive manufacturing methods by contributing to the design of novel structures optimized for additive manufacturing. Structures utilized in transportation, industry, medical equipment, and electronic components are susceptible to fatigue failure. There is a growing need to connect cutting-edge experimental characterization to numerical and artificially intelligent tools that are probabilistically grounded. Additive manufacturing has developed into a transformative digital manufacturing technique with a high degree of adaptability. The physics involved in this process chain are computationally too complex to grasp using conventional numerical techniques.

The present work focuses on the application of experimental techniques, numerical methods, and digital twinning to intelligently map fatigue strength to selective laser melting process parameters across a given process window. The experiments are designed to get a better knowledge of the relationship between process-induced microstructural characteristics and the ensuing consistency under laboratory and arbitrary conditions. A novel technique is developed in which the fatigue cyclic deformation and crack propagation behavior are digitally twined using a Bayesian-based machine learning algorithm. The suggested machine learning (ML) and Bayesian statistics techniques were used to construct the defect-correlated evaluation of fatigue strength. It facilitated the mapping of structural and process-induced fatigue features to a geometry-independent load density chart throughout a wide variety of fatigue regimes.

Dortmund   Prof. Dr.-Ing. Frank Walther
October 2022

Preface

This dissertation is the product of my work as a head of the Workgroup Modeling and Simulation at the Chair of Materials Test Engineering (WPT) of TU Dortmund University. At this point, I want to express my gratitude to everyone who contributed, directly or indirectly, to the realization of this work. I want to express my deepest gratitude to my doctor father, Prof. Dr.-Ing. Frank Walther, head of the Chair of Materials Test Engineering (WPT), for his ever-lasting support and guidance through technical assistance and motivating mindset. I warmly appreciate the interest of Prof. Dr.-Ing. Sebastian Münstermann in my work and his helpful and thoughtful comments. I want to express my gratitude to Prof. Dr.-Ing. A. Erman Tekkaya for his interest in this work and participation in the committee, and Prof. Dr. Moritz Schulze Darup for chairing the examination committee.

I am eternally grateful to my wife, Lobna Saeed, and my mother, Nahla Abdelkhalek, for their unconditional support throughout a journey spanning so many years. Not to be forgotten is the continuous support of my former colleague Mr. Shafaqat Siddique and current colleague, Mr. Mohamed Merghany. Mr. Shafaqat Siddique impacted my research residence and doctoral studies so positively and cooperated with me so closely, especially in the early phases. Mr. Jan Johannsen at Fraunhofer IAPT provided excellent support in specimen manufacturing and thermal measurements. I want to express my gratitude to the German Research Foundation (DFG) for funding the project entitled: "Mechanism-based understanding of functional grading focused on fatigue behavior of additively processed Ti-6Al-4V and Al-12Si alloys; project number 336368661." based on which this dissertation was prepared.

This work is dedicated to my wife, parents, and parents-in-law. I want to express my gratitude to them and my son Omar for understanding my continuous absence over the past several months and for their invaluable personal support.

Finally, I want to express my ever-lasting debt to my wife, Lobna, who has been standing fast, loving, and dependable help in every scenario. The prospect of our future together has always motivated me.

Dortmund   Mustafa Mamduh Mustafa Awd
October 2022

Abstract

Fatigue failure occurs in structures used in transportation, industry, medical equipment, and electronic components. There is an increasing need to build a link between cutting-edge experimental characterization and probabilistically grounded numerical and artificially intelligent tools. Additive manufacturing has evolved as game-changing digital manufacturing technology with high modularity. Researchers and scientists appear to be maximizing the yield of this technology by creating physics-based cause-and-effect relationships. The physics involved in this process chain is computationally prohibitive to comprehend using traditional computation methods. The mean surface temperature of the mechanical testing specimens exhibits a strong positive gradient up to a certain depth. The remelting procedure used successfully reduced the number of pores and total defect volume in Ti-6Al-4V. Scans using X-ray microcomputed tomography provide pure two-dimensional images in angular increments. X-ray microcomputed tomography was used to assess the efficacy of the remelting technique. Keyhole pores were formed as the scanning speed increased, accompanied by a decrease in metallurgical pores. Using machine learning and Bayesian statistics, a defect-correlated estimate of fatigue strength was developed. AlSi10Mg and Ti-6Al-4V had a tensile strength that was higher than the cast and wire+arc counterparts, respectively. Fatigue failures arose from the surface and sub-surface flaws in Ti-6Al-4V, whereas clusters of pores caused fatigue failures in most situations in AlSi10Mg. Since load is very low in the VHCF range, the crack propagation path showed linear elastic characteristics. Deviations could be due to microstructural discontinuities which exist in AlSi10Mg but not Ti-6Al-4V. The extended finite element method and contour integral techniques in Abaqus are used to study arbitrary crack paths. When the energy release rate exceeds a threshold value, the cracks will propagate from defects. Crack propagation rate curves are used to conclude about Paris constants of this studied specimen geometry in dependence on the applied load. Unforced fatigue cracks that originate from slip bands on the maximum shear planes are likely to adopt a mode II shearing failure that will branch into mode I tensile failure. The proposed technique for the simulation of cyclic deformation allows the identification of isotropic and kinematic hardening parameters, each on its own without interaction, since two independent procedures are used. A more complex cyclic deformation characteristic can be achieved by more pairs of back stress parameters. However, simulation efficiency will suffer from overfitting when an excessive number of back stresses is applied. Fatigue, which is a random variable, is studied in a Bayesian-based machine learning algorithm. In a Woehler field, the compatibility of the cumulative density function of the lifetime and the stress range must be ensured such that the minimization law of the weakest link principle is valid for Bayesian inference. The stress-life model was used based on the compatibility condition of life and load distributions. The defect-correlated assessment of fatigue strength was established using the proposed machine learning and Bayesian statistics algorithms. It enabled the mapping of structural and process-induced fatigue characteristics into a geometry-independent load density chart across a wide range of fatigue regimes.

Kurzfassung

Ermüdungsbrüche treten in Strukturen auf, die in der Verkehrstechnik, in der Fertigungstechnik, in medizinischen Geräten und in elektronischen Bauteilen verwendet werden. Es besteht ein zunehmender Bedarf eine Verbindung zwischen moderner experimenteller Charakterisierung und fundierten numerischen Werkzeugen sowie künstlicher Intelligenz herzustellen. Die additive Fertigung hat sich zu einer bahnbrechenden Fertigungstechnologie mit hoher Modularität entwickelt. Forscher/innen und Wissenschaftler/innen scheinen den Ertrag dieser Technologie zu maximieren, indem sie physikalisch begründete Ursache-Wirkungs-Beziehungen schaffen. Die Physik, die dieser Prozesskette zugrunde liegt, ist mit herkömmlichen Berechnungsmethoden nicht zu erfassen. Die mittlere Oberflächentemperatur der gefertigten, mechanischen Prüfkörper weist bis zu einer bestimmten Tiefe einen starken positiven Gradienten auf. Das angewandte Umschmelzverfahren reduziert erfolgreich die Anzahl der Poren und das Gesamtdefektvolumen in TiAl6V4. Die Mikrocomputertomographie liefert reine zweidimensionale Bilder in Winkelinkrementen. Sie wurde verwendet, um die Wirksamkeit des Umschelzverfahren zu beurteilen. Mit zunehmender Scangeschwindigkeit treten sog. „keyhole" Poren auf, während die metallurgischen Poren abnahmen. Mithilfe von maschinellem Lernen und Bayes'scher Statistik wurde eine defektkorrelierte Schätzung der Ermüdungsfestigkeit entwickelt. AlSi10Mg und TiAl6V4 wiesen eine höhere Zugfestigkeit auf als die gegossenen bzw. die mit Lichtbogenschweißen hergestellten Gegenstücke. Ermüdungsbrüche sind bei TiAl6V4 auf oberflächennahe und Oberflächenfehler zurückzuführen, während bei AlSi10Mg in den meisten Fällen Porenbündel den Ermüdungsbruch verursachen. Da die Belastung im VHCF-Bereich sehr gering ist, zeigt der Rissausbreitungsweg linear-elastische Eigenschaften. Dies könnte auf mikrostrukturelle Diskontinuitäten zurückzuführen sein, die in AlSi10Mg, aber nicht in TiAl6V4 vorhanden sind. Die erweiterte Finite-Elemente-Methode und die Konturintegraltechnik in Abaqus können zur Untersuchung beliebiger Risspfade verwendet werden. Wenn die Energiefreisetzungsrate einen Schwellenwert überschreitet, breiten sich die Risse von den Defekten ausgehend aus. Mit Hilfe von Risswachstumskurven lassen sich Rückschlüsse auf die Konstanten des Paris-Gesetzes für die untersuchte Proben-geometrie in Abhängigkeit der aufgebrachten Last ziehen. Ermüdungsrisse, die von Gleitbändern auf den Hauptgleitebenen ausgehen, werden voraussichtlich zu einem Scherversagen (Mode II) führen, das in ein Zugversagen (Mode I) übergeht. Die vorgeschlagene Technik zur Simulation der zyklischen Verformung ermöglicht die Identifizierung isotroper und kinematischer Verfestigungsparameter ohne Wechselwirkung, da zwei unabhängige Verfahren verwendet werden. Eine komplexere zyklische Verformungscharakteristik kann durch mehr Paare von Rückspannungsparametern erreicht werden. Allerdings kann die Simulationseffizienz unter einer Überanpassung leiden, wenn eine zu große Anzahl von Rückspannungen angewendet wird. Die Ermüdung ist eine Zufallsvariable, die mit einem Bayes-basierten maschinellen Lernalgorithmus untersucht wird. In einer Gruppe Wöhler-Kurven muss die Kompatibilität der kumulativen Dichtefunktion der Lebensdauer und des Beanspruchungsbereichs gewährleistet sein, sodass das Minimierungsgesetz des Schwächsten-Glied-Prinzips für die Bayes'sche Inferenz gilt. Das Spannung-Lebensdauer-Modell wurde auf der Grundlage der Kompatibilitätsbedingung von Lebensdauer- und Lastverteilungen verwendet. Die defektkorrelierte Bewertung der Ermüdungsfestigkeit wurde mithilfe der vorgeschlagenen Algorithmen für maschinelles Lernen und Bayes'scher Statistik erstellt. Sie ermöglichte die Abbildung von strukturellen und prozessinduzierten Ermüdungsmerkmalen in ein geometrieunabhängiges Lastdichtendiagramm über einen weiten Bereich von Ermüdungszuständen.

Contents

Abbreviations

μ-CT	X-ray microcomputed tomography
2D	Two dimensional
3D	Three dimensional
AB	As-built
AE	Acoustic emission
AF	Armstrong-Frederick
AI	Artificial intelligence
Al	Aluminum
AM	Additive manufacturing
ANN	Artificial neural network
ASTM	American society for testing and materials
BCC	Body-centered cubic
BN	Bayesian network
CAD	Computer-aided design
CAE	Computer-aided engineering
CAT	Constant amplitude test
CDF	Cumulative distribution function
CDM	Continuum damage mechanics
CG	Conventional grain
CNC	Computer numerical control
COD	Crack opening displacement
CP	Cyclic plasticity
CPFEM	Crystal plasticity finite element method
CPZ	Cyclic plastic zone
CT	Compact tension
DFR	Detail fatigue rating
DIN	Deutsches Institut für Normung
DLD	Direct laser deposited

DMLS	Direct metal laser sintering
DOF	Degree of freedom
DP	Dual-phase
DRV	Dynamic recovery
DRX	Dynamic recrystallization
DVC	Digital volume correlation
EBM	Electron beam melted
EBSD	Electron backscattered diffraction
EDS	Energy-dispersive X-ray spectroscopy
EDX	Energy-dispersive X-ray
EIFS	Equivalent initial flaw size
EN	European Norm
Ene.	Energy
EPFM	Elastic plastic fracture mechanics
FCC	Face centered cubic
FCG	Fatigue crack growth
FCGR	Fatigue crack growth rate
FDM	Fused deposition modeling
FE	Finite element
Fe	Iron
FEM	Finite element method
FLD	Forming limit diagram
fps	Frames per second
FS	Fatemi-Socie
GFEM	Generalized finite element method
GUI	Graphical user interface
HCF	High cycle fatigue
HCP	Hexagonal close-packed
HIP	Hot isostatic pressing
HMT	Hidden Markov tree
HRC	Rockwell hardness tester scale
HV	Vickers hardness test number
IR	Infrared
ISO	International organization for standardization
LCF	Low cycle fatigue
LEFM	Linear elastic fracture mechanics
LENS	Laser engineered net shaping
LIT	Load increase test
LOF	Lack of fusion

LS	Least-squares
LSM	Level set method
Maxe	Maximum nominal strain criterion
Maxpe	Maximum principal strain criterion
Maxps	Maximum principal stress criterion
Maxs	Maximum nominal stress criterion
MERR	Maximum energy release rate
Mg	Magnesium
ML	Machine learning
MLE	Maximum likelihood estimation
MPZ	Monotonic plastic zone
MSC	Microstructural short crack
MTS	Maximum tangential stress criterion
NN	Neural network
ODB	Output database
PBF	Powder bed fusion
PDF	Probability density function
PH	Platform heating
PHILSM	Signed-distance function
PM	Powder metallurgy
PPNN	Probabilistic physics-guided neural network
PSB	Persistent slip bands
PWM	Probability-weighted moments
Px	Projected area in x-direction
Py	Projected area in y-direction
Pz	Projected area in z-direction
Quade	Quadratic nominal strain criterion
Quads	Quadratic nominal stress criterion
RBF-NN	Radial basis function neural network
RFR	Random forest regression
ROI	Region of interest
RVE	Representative volume element
Sc	Scandium
SED	Strain energy density
SEM	Scanning electron microscopy
Si	Silicon
SIF	Stress intensity factor
SLM	Selective laser melting
SLS	Selective laser sintering

S-N	Woehler curve
SR	Stress relief
SSY	Small scale yielding
STEP	Standard for exchange of product model data
STL	Standard tessellation language
SVM	Support vector machine
SWT	Smith-Watson-Topper
T. F.	Stress triaxiality factor
TEM	Transmission electron microscopy
Ti	Titanium
TM	Tanaka-Mura
UFG	Ultra-fine grained
UMAT	User material
USF	Ultrasonic fatigue
UTS	Ultimate tensile strength
V	Vanadium
VCCT	Virtual crack closure technique
VHCF	Very high cycle fatigue
WAAM	Wire+arc additive manufacturing
XFEM	Extended finite element method
XRD	X-ray diffraction

List of symbols

Latin symbols

$\vec{a}_I$	Nodal enriched degree of freedom
$\vec{b}_I^\alpha$	Nodal degree of freedom
n_c/N_H	Consumed life ratio
n_r/N_L	Remaining life ratio
$\vec{u}_I$	Nodal displacements (mm)
A_0	Initial amplitude at time t_0
A_c	Projected contact area (mm^2)
a_N	Crack length development in N cycles (mm)
b_0	Torsional fatigue strength exponent
c_0	Torsional fatigue ductility exponent
D^α	Drag stress (MPa)
E^*	Modulus of reduction (GPa)
E_c^*	Reduced contact modulus (GPa)
E_{dyn}	Dynamic elastic modulus (GPa)
E_{elpl}	Elastoplastic stiffness matrix
E_f	Modulus of the surface (GPa)
E_i^*	Modulus of reduction indenter (GPa)
E_{ind}	Elastic modulus of the indenter (GPa)
$E_{kin}(t)$	Kinematic energy (J)
$E_{pot}(t)$	Potential energy (J)
E_s	Modulus of the subsurface (GPa)
E^{SI}	Specimen + indenter modulus (GPa)
f_{max}	Maximum indentation force (mN)
f_{min}	Minimum indentation force (mN)
df/dt	Force application rate (mN/s)

f_{tol}	Crack propagation tolerance
$F_\alpha(x)$	Crack tip asymptotic function
G_c	Critical energy release rate (N/mm)
G_C	Fracture energy release rate (N/mm)
G_{equiv}	Equivalent energy release rate (N/mm)
G_{equivc}	Critical equivalent energy release rate (N/mm)
G_I	Mode I energy release rate (N/mm)
G_{Ic}	Mode I critical energy release rate (N/mm)
G_{II}	Mode II energy release rate (N/mm)
G_{IIc}	Mode II critical energy release rate (N/mm)
G_{III}	Mode III energy release rate (N/mm)
G_{IIIc}	Mode III critical energy release rate (N/mm)
G_{max}	Maximum energy release rate (N/mm)
G_{pl}	Energy release rate upper limit (N/mm)
G_s	Global seed
G_{thresh}	Energy release rate threshold (N/mm)
$H(x)$	Heaviside distribution-following jump enrichment function
h_c	Contact depth (μm)
h_{max}	Maximum indentation depth (μm)
I_0	Frequency of checking the iteration residuals
I_0	Set of runouts
I_1	Set of non-runouts
I_A	Maximum number of cutbacks permitted in an increment
I_R	Staring point of the convergence checks
J_{Ic}	Mode I critical contour integral value (N/mm)
K_c	Critical stress intensity factor (MPa$\sqrt{}$mm)
K_{IC}	Mode I fracture toughness (MPa$\sqrt{}$mm)
K_{II}	Mode II stress intensity factor (MPa$\sqrt{}$mm)
K_{max}	Maximum stress intensity factor (MPa$\sqrt{}$mm)
K_{th}	Stress intensity factor threshold (MPa$\sqrt{}$mm)
N_f^a	Expected fatigue lifetime (cycles)
N_f	Number of cycles of failure (cycles)
$N_I(X)$	Nodal shape function
S_f	Failure stress (MPa)
T_i	Stress vector (MPa)
u_i	Displacement vector (mm)
w_f	Fracture energy (J)
W_s	Work needed to create the crack surfaces (J)

$\boldsymbol{x'}$	Deviatoric back stress tensor (MPa)
K^{α}	Length scale-dependent threshold stress (MPa)
ΔK_I	Mode I stress intensity factor range (MPa$\sqrt{}$mm)
ΔK	Stress intensity factor range (MPa$\sqrt{}$mm)
a	Fatigue crack length (mm)
$2a$	Crack length in an infinite plate (mm)
A	Notch contact area, Area of the crack, Excitation amplitude (mm^2, μm)
B	Thickness of the compact specimen, Bias of fatigue lifetime threshold (mm, cycles)
b	Rate of change of yield surface, Axial fatigue strength exponent
C	Fatigue crack propagation rate coefficient, Kinematic hardening parameter, Bias of fatigue endurance limit
c	Axial fatigue ductility exponent
D	Spot size (mm)
da/dN	Crack propagation rate (mm/cycle)
dp	Cumulative plastic strain
dR	Rate of change of the isotropic hardening
ds	Increment along the path
E	Elastic modulus (GPa)
Ev	Energy density (J/mm^3)
f	Testing frequency (Hz)
F	Yield surface
$F(N)$	Failure probability
G	Energy release rate, Shear Modulus, A suitable rising function of lifetime (N/mm, GPa, -)
h_t	Hatching distance (mm)
H	Hardening modulus (GPa)
h	Reading depth of the indenter (μm)
$H(x)$	Heavy side function
I	Intensity of photons (W/sr)
$\boldsymbol{I}$	Second-order identity tensor
J	Contour integral (N/mm)
K	Stress intensity factor (MPa$\sqrt{}$mm)
k_s	Stiffness (N/mm)
k	Initial size of the yield surface, Fatemi-Socie damage parameter (MPa, -)
$\mathbb{L}$	Trajectory (mm)
l	Length (mm)

L_c	Characteristic length (mm)
L	Specimen size effect
m	Mass, Fatigue crack propagation rate exponent, Number of macroscale back stress components, Crystal flow exponent (Kg, -)
M	A probability density function of microstructure
$\mathbb{M}$	Sample of microstructure
N	Number of cycles (cycles)
P	Power, Load applied by the indenter (W, mN)
p	Equivalent plastic strain, Percentile
P	A probability density function of porosity
Q	Maximum change in the size of the yield surface
R	Load/stress ratio, Isotropic hardening variable
$\mathbb{R}$	Domain of real numbers
S	Slope of the unloading curve
$S\text{-}N$	Woehler curve
T	Temperature (°C, K)
t	Layer thickness (mm)
vs	Scanning speed (mm/s)
w	Strain energy (J)
W	Width of the compact-tension specimen (mm)
x	Back stress tensor (MPa)
ΔG_{th}	Energy release rate range threshold (N/mm)
ΔK_{th}	Stress intensity factor range threshold (MPa$\sqrt{}$mm)
ΔF	Applied force range (kN)
ΔG	Energy release rate range (N/mm)
$Q(N, \Delta\sigma)$	Probability distribution of lifetime in dependence of applied stress range

Greek symbols

$\dot{\gamma}^\alpha$	Slip system shearing rate (s^{-1})
$\dot{\gamma}_0$	Reference shear rate (s^{-1})
v^e	Elastic Poisson's ratio
v_i	Poisson's ratio of indenter
v^p	Plastic Poisson's ratio
α^{th}	Slip system
γ_f'	Torsional fatigue ductility coefficient

γ_{max}	Maximum shear strain
ε_0	Initial strain amplitude
$\varepsilon_{a,f}$	Fracture strain amplitude
$\varepsilon_{a,p}$	Plastic strain amplitude
ε_p	Quasi-static plastic strain
$\varepsilon_{a,t}$	Total strain amplitude
ε^{el}	Constitutive elastic strain
ε'_f	Axial fatigue ductility coefficient
ε_f	Fracture strain
$\varepsilon_{m,p}$	Plastic mean strain
$\varepsilon_{m,t}$	Total mean strain
ε_n	Strain in pure normal mode
ε^{pl}	Constitutive plastic strain
ε_s	Strain in first shear direction
ε_S	Stable strain amplitude
ε_t	Total strain
ε_{ts}	Strain in second shear direction
$\hat{\theta}$	Reference angle of the first fracture plane (rad)
σ'	Deviatoric stress tensor (MPa)
σ^2	Variance (unit^2)
$\sigma_{a,f}$	Fracture stress amplitude (MPa)
$\sigma_{a,start}$	Starting stress amplitude (MPa)
σ_a	Stress amplitude (MPa)
σ'_f	Axial fatigue strength coefficient (MPa)
σ_m	Mean stress (MPa)
σ_{max}	Maximum stress (MPa)
σ_n^{max}	Maximum normal stress (MPa)
σ_{max}^0	Maximum stress at the first loop (MPa)
σ_{max}^s	Maximum stress at the stable loop (MPa)
σ_n	Stress in pure normal mode (MPa)
σ_s	Stress in first shear direction (MPa)
σ_{ts}	Stress in second shear direction (MPa)
σ_y	Yield strength (MPa)
σ_{y0}	Initial yield strength (MPa)
σ_h	Hydrostatic stress (MPa)

σ_{eq}	von Mises stress (MPa)
τ_f'	Torsional fatigue strength coefficient (MPa)
τ^α	Resolved shear stress (MPa)
χ^α	Crystal back stress (MPa)
$\Delta\sigma$	Stress range (MPa)
$d\varepsilon$	Rate of strain (s^{-1})
$d\lambda$	Constancy parameter
$Tr(\sigma)$	Trace operator of a tensor
υ	Poisson's ratio
α	Alpha phase, Gamma distribution parameter
α'	Acicular martensite
β	Beta phase, First shape parameter
β'	Mg2Si precipitates
γ	Kinematic hardening parameter, Second shape parameter
$\Gamma(\alpha)$	Gamma distribution
δ	Scale parameter
ε	Strain
$\in$	Geometry factor of the indenter
η	Exponent for BK law and Reeder law
θ	Equivalent half-angle at the vertex of the tip (rad)
λ	Location parameter
μ	Linear attenuation coefficient (m^{-1})
μ	Mean
σ	Stress (MPa)
τ	Shear stress (MPa)
$\varphi(x, t)$	Level set method function
ϕ	Phase shift
ω	Angular frequency (rad/s)
Γ	Path of the contour integral (mm)
Π	Fracture potential energy (J)

List of Figures

List of Tables

Introduction and Objectives

Manufacturing firms in high-wage nations like Germany are becoming increasingly vulnerable to global competition [1]. These businesses are being challenged by rising demand for customized, high-quality, and low-cost products. In the last few years, additive manufacturing (AM) technologies have increased their technological capabilities and now hold tremendous promise for addressing these challenges. On the other hand, the theory of metal fatigue has been studied for almost 160 years, but there are still many unsolved concerns about the phenomenon especially regarding AM structural integrity.

Figure 1.1 explains the locations in the body of a commercial airliner where fatigue resistance is a structural requirement [2]. The inhomogeneities in metallic structures are one of the most cited reasons for metal fatigue in industrial and engineering structures. This topic became of extreme importance to researchers and in the industry as well because of the severity of problems that occur if fatigue behavior is not well studied. The problem here is that fatigue testing is extremely expensive and time-consuming relative to other mechanical tests, like the tensile test, for example. Therefore, much research work is being directed to find methods of simulating and predicting the fatigue behavior of materials and parts [3]. Recent technological advancements enable the fabrication of new metallic materials with better fatigue characteristics via microstructure tailoring. Considering these encouraging discoveries, there is an increasing need to build a link between cutting-edge experimental characterization and physically grounded numerical and artificially intelligent tools. In the early twentieth century, understanding of fatigue was still in its infancy, particularly in terms of the material

© The Author(s), under exclusive license to Springer Fachmedien Wiesbaden GmbH, part of Springer Nature 2022
M. Mamduh Mustafa Awd, *Machine Learning Algorithm for Fatigue Fields in Additive Manufacturing*, Werkstofftechnische Berichte | Reports of Materials Science and Engineering, https:doi.org/10.1007/978-3-658-40237-2_1

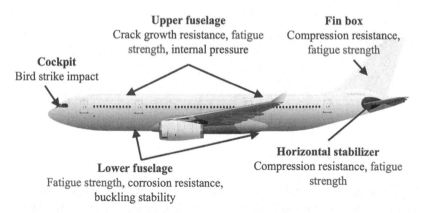

Upper fuselage
Crack growth resistance, fatigue
strength, internal pressure

Fin box
Compression resistance,
fatigue strength

Cockpit
Bird strike impact

Lower fuselage
Fatigue strength, corrosion resistance,
buckling stability

Horizontal stabilizer
Compression resistance, fatigue
strength

Figure 1.1 Distribution of desired mechanical properties across the structure of a commercial airliner [2]

and engineering elements of fatigue. For a long period of time, fatigue has piqued scientific curiosity. It is essential to have a historical understanding of fatigue for the present generation of scientists and researchers to expand on that knowledge. Fatigue failure occurs in structures used in transportation (see Figure 1.1), industry, medical equipment, and electronic components.

The objective of fatigue design is to avoid fatigue problems for safety, economy, durability, and liability reasons. Recent developments in modeling and simulation, as well as in situ experimental approaches, have aided in our knowledge of the initiation and early growth phases of fatigue fractures. Engineers and materials scientists have long attempted to establish a relationship between the detailed microstructures of advanced engineering alloys and the material's performance. Microstructure influences high cycle fatigue resistance and contributes to the fatigue response's heterogeneity. Additive manufacturing (AM) has evolved as game-changing digital manufacturing technology with high modularity. Researchers and scientists appear to be maximizing the yield of this technology by creating physics-based cause-and-effect linkages. However, the physics involved in this process chain is computationally prohibitive to comprehend using traditional computation methods. Due to its effectiveness in classification, regression, and clustering, machine learning (ML) has garnered growing interest in AM. ML will be used to generate novel high-performance metamaterials and optimal topological designs in design for additive manufacturing. However, a balance must be achieved between bias and variance by

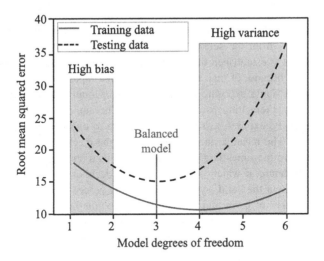

Figure 1.2 Fine-tuning machine learning models to achieve a balanced trade-off between bias and variance [4]

optimizing the number of degrees of freedom, according to Figure 1.2 [4]. There has been a growing concern regarding data security in AM, as data breaches can arise as a result of machine learning algorithms; however, the potential outweighs the risks [5].

With the accumulated knowledge in the background in chapter 2, it is possible to assess the significance of realistic simulation testing that bridges the gaps between modern experimental techniques and artificial intelligence algorithms. However, it is necessary to examine issues that should be explored further, such as the role artificial intelligence can contribute to the fatigue problem. Stress analysis does not have to be a significant impediment; rather, it is the fatigue behavior of structural parts that cannot be accurately represented in sufficient depth and the relation to multiscale stochastic structural features. That will be the aim of this work that centralizes around the objective of creating digital twins based on experimental data according to the view of industry 4.0 [6]. The output of the two digital twins, which are a crack propagation twin, and a cyclic deformation twin, will be used in a probabilistic machine learning algorithm. The algorithm will establish a fatigue property map for functionally graded AlSi10Mg and Ti-6Al-4V based on the selective laser melting process.

In chapter 3, the calibration and training data are established through selective laser melting and multiscale characterization. This work uses two lightweight alloys from the selective laser melting material palette: AlSi10Mg and Ti-6Al-4V. It is established that turbulent melt pools collapse over pressure-decreasing voids because of melt pool instabilities and rapid cooling rates. When compared to the single exposure mean surface temperature, heat build-up was significant in most remelting parameter choices. The author proposes that current build layers be exposed to secondary exposure to enhance degassing prior to solidification. The mean surface temperature of the mechanical testing specimens, which were ten times larger in height, exhibits a strong positive gradient up to a certain depth, at which point the gradient declines and the temperature stabilizes throughout the build layers. The remelting procedure used successfully reduced the number of pores and the total defect volume in Ti-6Al-4V. To evaluate their effect on the formation of porosity in the transition zone between two processing conditions, specimens with three distinct vector overlaps/gaps should be generated.

In chapter 4, the stress intensity factor (SIF) at the crack tip will be computed using Abaqus's extended finite element method (XFEM) and contour integral approaches. To investigate the fatigue crack development process in additively built fatigue specimens, XFEM was utilized to mimic the crack propagation process using an arbitrary fracture path. The effect of loading level and testing frequency on fatigue lifetime was investigated using crack propagation rate curves. Scans using X-ray microcomputed tomography (μ-CT) provide pure two-dimensional (2D) images in angular increments. Numerous reductions and simplification approaches must be used to minimize the number of faces and vertices. FreeCAD was used to separate the cylindrical shell from the internal porosities and transform μ-CT scans into finite element models. All simulations were conducted on a cylindrical fatigue specimen that was fabricated via selective laser melting (SLM). The model for predicting the fatigue lifetime based on Fatemi-Socie damage is proportional to the greatest shear strain created during cyclic loading simulations. The strain range of this load level is determined using the outcome of a uniaxial fatigue test simulation. Segmentation and discretization of the cyclic stress-strain response in the load increase test (LIT) were proven to be helpful in optimizing back stress pairings in AlSi10Mg and Ti-6Al-4V across a wide load range. The bias in the simulated and calibrated values is due to the experimental data for both alloys being statistically distributed. Having an excessively specified material calibration does not result in increased accuracy and might result in an inefficient simulation. Unforced fatigue fractures originating from slip bands on the maximum shear planes are likely to break into a mode II shearing mode before branching into a mode I tensile mode.

In chapter 5, within a specially created machine learning method, fatigue is handled as a random variable regardless of whether it is studied in the laboratory or on-site. A three-parameter Weibull distribution was used to represent the fatigue strength in the S-N curves for a stress-controlled statistical fatigue lifetime model. When one fatigue-inducing parameter must be mapped onto the fatigue lifetime, the joint probability function will be solved using a unique transformation matrix referred to as the Jacobian of the mapping. Machine learning (ML) and Bayesian statistics techniques were used to develop a defect-correlated estimate of fatigue strength. Cumulative energy absorption rises in proportion to the power of the energy function. The increased energy density in SLM causes the Ti-6Al-4V powder bed to over-melt. Additional impacts such as residual stress mapping on fatigue strength will be included in future investigations following this work.

Background on Process-Property Relationship

2

2.1 Fatigue Strength of Additively Processed Al and Ti Alloys

Additive manufacturing (AM) is an advanced manufacturing technology enabling the production of complex shapes by adding material layer-upon-layer [7]. AM has come a long way since its start in the 1980s. AM is a well-known method of producing three-dimensional objects. Powder bed fusion (PBF)-based AM has received special interest since it is cost-effective, eco-friendly, and adaptable to flexible powder materials [8]. However, AM's potential for design flexibility is limited when producing complex shapes, see Figure 2.1 [9]. Support structure research is important for additive manufacturing's future and advancement [10]. Regarding mechanical properties, Ti-6Al-4V alloy, Al-Si alloys, and 316L steel have been extensively investigated [11].

A fundamental technological obstacle preventing enterprises from adopting AM technologies is a lack of quality assurance with AM parts. Researchers have looked into new types of instrumentation and adaptive approaches that, when combined, will improve quality assurance even further [12]. It is suggested that qualification techniques and documentation be improved, and areas with substantial potential for future research are investigated [13]. Biotechnology is being transformed by additive manufacturing, and architects are finding it easier to do their work. Geometrical characteristics are extracted from a computer-aided design (CAD) file and translated to a stereolithography standard tessellation language (STL) file in additive manufacturing methods [14]. AM metallic materials have static mechanical properties that can compete with those of traditionally manufactured metallic components. Feature qualifications have also been recognized as a barrier to wider deployment of AM in precision industries, see

© The Author(s), under exclusive license to Springer Fachmedien Wiesbaden GmbH, part of Springer Nature 2022
M. Mamduh Mustafa Awd, *Machine Learning Algorithm for Fatigue Fields in Additive Manufacturing*, Werkstofftechnische Berichte | Reports of Materials Science and Engineering, https:doi.org/10.1007/978-3-658-40237-2_2

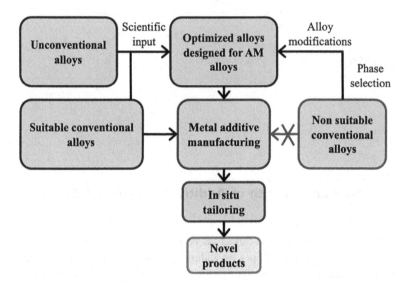

Figure 2.1 Architectural flow of suitability of alloy compositions for metal additive manufacturing according to Zhang et al. [9]

Figure 2.2 [15]. Selective laser melting (SLM) allows you to customize the mechanical attributes of your items to a great extent, but it is still in need of substantial improvements. It was simply adapted to a variety of materials which paves the way for better-functioning medical implants and instruments, as well as a broader range of machine parts [16].

Figure 2.2 Relationship between beam power, deposition rate, and feature quality with respect to building rate and feature definition according to Frazier [15]

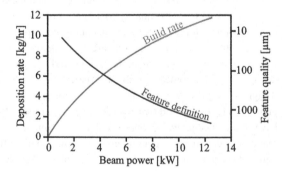

The SLM process chain relies on process monitoring and real-time process management to improve part quality and repeatability. The quality of metallic AM parts is currently insufficient to meet the serial production rigorous criteria. To solve this obstacle, more basic and applied research is required [17]. A basic advantage of various AM systems is their ability to process different materials [18]. The fatigue behavior of metallic materials generated using additive fabrication is unclear, despite several experimental efforts undertaken in the past few years. The fatigue characteristics and crucial variables of components manufactured utilizing conventional production procedures were found to be quite similar. These findings demonstrated that defect-tolerant design concepts might be used for additively manufactured components as well [19].

2.1.1 Aluminum Alloys (Al)

The effect of platform heating (PH) and post-process stress relief on component characteristics such as process-induced flaws, which are essential for fatigue loading, is investigated in a study for AlSi12 alloy. There have been investigations on extremely high cycle fatigue as well as fatigue crack development. By heating the platform, improved fatigue and crack development behavior is demonstrated, as well as the fatigue crack nucleation mechanisms, see Figure 2.3. Reduced temperature gradients caused by base plate heating contribute to increased fatigue reliability by decreasing fatigue crack initiation caused by material flaws [20].

Figure 2.3 Very high cycle fatigue strength under the influence of platform heating of selective laser melted AlSi12 according to Siddique et al. [20]

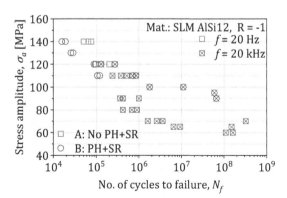

As a result of fatigue loading, defects were the primary cause of fracture initiation. This dispersion in strength was caused by interactions between defect size, location, load magnitude, and type. Weibull's probability density function was used to interpret a specimen's fatigue life [21]. Fatigue characterization was conducted at frequencies of 20 Hz and 20 kHz for high cycle fatigue (HCF) and for very high cycle fatigue (VHCF) up to 1E9 cycles, respectively. The results indicate that SLM components outperform cast materials. However, for dependable fatigue performance, microstructural characteristics, as well as process-induced irregularities, must be controlled [22]. A study examined AlSi10Mg specimens generated by SLM. Microstructural investigations indicated that the process parameters permitted the production of a homogeneous microstructure. Large Gaussian specimens were ultrasonically tested if they failed until $6 \cdot 10^8$ cycles [23]. The fatigue life was shown to be related to the size of the crack-initiating pore on the fracture surface. Using extreme-value theory, the pore size distribution was utilized to estimate fatigue life; a good agreement was achieved [24].

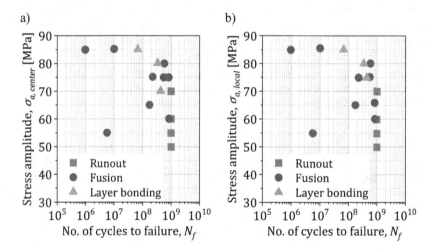

Figure 2.4 Very high cycle fatigue strength of AlSi10Mg according to Paolino et al. [25]

For experimentally evaluated VHCF response of SLM AlSi10Mg specimens constructed in the vertical direction, the quasi-static mechanical characteristics were bigger than and close to those reported in the literature for cast alloys [26]. The surface roughness is critical for vertically constructed AlSi10Mg components' VHCF loading. Removal of a surface material layer does not impact the average VHCF strength substantially (Figure 2.4), but it allows for lowering the experimental scatter and hence improve the VHCF strength with the same reliability required [25]. With a run-out load of 200 MPa, SLM Al-Si alloys had the best fatigue performance. The material's ductility and fatigue characteristics vary according to the type of defect. However, the various flaws had no effect on the tensile strength while ductility remained constant [27]. The most frequently occurring flaws in specimens were pores and clusters of pores. When VHCF specimens were annealed at 244 °C, the VHCF material reactions were enhanced. Residual stress minimization is unsuccessful if the thermal history of the microstructure is fundamentally altered by the heat treatment. Reduced residual stresses had little effect due to the predominance of microstructural changes [28].

Surface treatments have been shown to improve the fatigue performance of SLM specimens considerably. Additionally, they have the potential to reduce the need for polishing procedures that are now done on as-built material [29]. The impact on the fatigue resistance of AlSi10Mg additively produced specimens after shot-peening surface treatment was positive. Nano-indentation studies indicated shot-peening produced surface hardening, with the surface-to-inside hardness profile vanishing 50 μm below the surface. The final die-cast fracture region revealed a fracture surface including many cleavage facets and microcracks [30]. Rotational bending fatigue tests were performed on cast aluminum alloy A356 specimens, see Figure 2.5. The experimental fatigue limit at the high cooling rate specimens was lower than predicted [31].

Figure 2.5 Influence of cooling rate (A-C highest to lowest) on fatigue strength of cast A356 cast Al alloy according to Tajiri et al. [31]

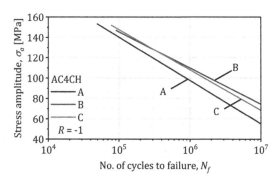

After peak hardening to T6 treatment, the microstructure is homogenous, with dendrites, laser traces, and heat-impacted zones dissolved and Si-particles assuming a globular form. Peak hardening has the biggest influence on fatigue resistance, whereas platform heating has the smallest effect, see Figure 2.6 [32]. The linear area of the crack propagation rate da/dN vs. stress intensity factor range ΔK graphs of the AlSi10Mg alloy generated by SLM were investigated for the development of high cycle fatigue cracks. The fatigue crack growth (FCG) curves demonstrated that heat treatment had a positive effect on the FCG curves. On the other hand, the development of fatigue cracks is largely reliant on the load orientation of the specimen [33].

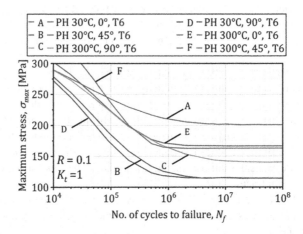

Figure 2.6 Influence of peak hardening of fatigue strength of AlSi10Mg according to Brandl et al. [32]

Controlling defects

The defect-tolerant design combined with newly established fracture mechanics is a key to expressing fatigue strength and material quality [34]. For defect-tolerant design, an expanded Kitagawa-Takahashi fatigue diagram was created that contains a traditional safe-life zone and defect-determined finite life region [35]. Surface or subsurface gas pores caused fatigue failure. Gas pores with a percentage of 0.2—1.6 % and a diameter of 20—55 μm are found in SLM-produced Al-Si alloy components [36]. Specimens free of contouring defects exhibited a smoother surface and higher compressive residual stresses in the load direction, which might explain their higher sustained cycles to failure [37]. The 3D pore geometry was shown to be crucial in determining the fatigue limit. This was done using a finite element model of actual three-dimensional (3D) pores [38]. Despite shortcomings, high-quality process parameters allowed the development of components with a fatigue life at least similar to those of materials created using conventional manufacturing procedures [39].

Heat treatment
Heat treatment changes the size and morphology of Si, which leads to a decrease in fatigue properties [40]. The primary difference between SLM and conventional materials in terms of fatigue behavior is not only in terms of overall fatigue life but also in terms of the duration of fatigue damage stages and the process of fracture formation. The highest fatigue resistance was achieved using fresh powder in a single construction operation [41]. After different heat treatments, the fatigue limit of the machined and polished specimens was the greatest (failure stress $S_f = 125$ MPa, $N_f = 10^7$ cycles). The yield strength, hardness, and fatigue limit are all reduced throughout the stress relief (SR) and hot isostatic pressing (HIP) cycles [42]. Machining heated-treated SLM AlSi10Mg did not result in an increase in fatigue life at higher stress levels [43].

2.1.2 Titanium Alloys (Ti)

Sensitivity to defects
Despite the huge potential of Ti alloys within AM technology, reliable mechanical properties are a precondition for future serial production [44]. SLM defect features could not be identified in X-ray microcomputed tomography (μ-CT) scans because they were small, while electron beam melted (EBM) defect features were large enough to be visible, however [45]. SLM-made parts are less fatigue-resistant than conventionally processed components. μ-CT scans, metallographic, and fractographic inspections supplement residual stress and microhardness measurements, provided an interpretative model that incorporates post-processing adjustments, see Figure 2.7 [46]. The Basquin equation was used to predict the fatigue life of Ti-6Al-4V parts subjected to fatigue loading. The model was created by comparing X-ray μ-CT data and consideration of idealized pore geometries. This stress amplitude was utilized in the Basquin equations for fatigue life prediction [47].

An investigation was carried out using a conventional linear elastic fracture mechanics (LEFM) approach and data from tests on compact-tension (CT) specimens that were also in their as-manufactured condition with a variety of build directions. Cracks were typically initiated at lack of fusion (LOF) flaws, reducing

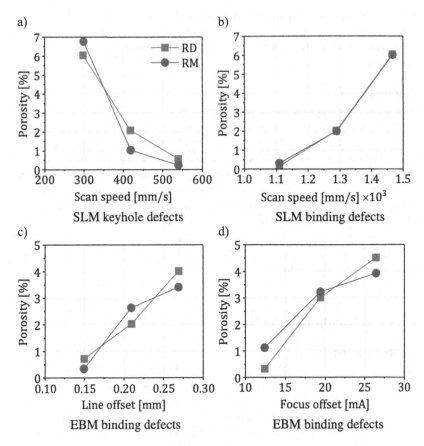

Figure 2.7 Influence of process parameters on porosity levels (Porosity estimate: RD ➜ micrographs; RM ➜ μ-CT) according to Gong et al. [45]

the total fatigue life [48]. Roughness reduction resulted in a significant increase in fatigue strength, from 300 to 775 MPa (after $3\cdot10^7$ cycles) [49]. Increased fatigue endurance limit was attributed to the elimination of internal porosity [50]. SLM-processed Ti-6Al-4V has a better monotonic stress-strain behavior than its traditionally produced counterparts. This is not the case, however, for high-cycle fatigue tests due to process-induced defects generated by SLM [51]. The most effective measures for enhancing high and very high cycle fatigue resistance, respectively, were determined to be shot peening and HIPing [52]. A study was carried out by scientists at the German Aerospace Center. The goal was to restore Ti-6Al-4V's ductility and HCF strength to its conventional levels, see Figure 2.8. Increased scanning speed resulted in a rapid reduction in porosity, whereas increased scanning velocity resulted in increased porosity due to a significant number of small pores and surface roughness (< 4 μm) [53]. The HCF experiments established that polished SLM Ti-6Al-4V has superior fatigue properties, with an endurance limit of 500 MPa, see Figure 2.9. Although the material's endurance limit was reduced in the as-built condition due to the process's intrinsic high surface roughness [54], the fatigue properties appear to be suitable for a broad variety of applications in the aviation and medical industries [55]. Even pores under a few microns of diameter proved to be fatal based on further characteristics. Fatigue is more sensitive to the size and shape of internal pores than to their location across several AM processes [56, 57].

Figure 2.8 Suppression of process-induced defects by hot isostatic pressing (HIP) to improve the fatigue strength of Ti-6Al-4V according to Kasperovich and Hausmann [53]

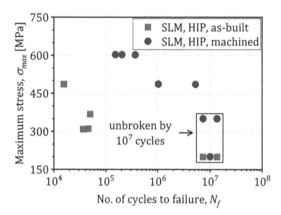

Influence of microstructure

Even without microstructure modifications, stress-reducing heat treatment at lower temperature results in a fatigue fracture development behavior equivalent to that of conventional Ti-6Al-4V [58]. In the as-built condition, there are substantial differences between SLM and cold-rolled Ti-6Al-4V in fatigue and microstructure characteristics [59]. HIP post-treatment removes intrinsic process defects and substantially enhances fatigue performance through recrystallized microstructure [60]. Despite coarsened microstructure and reduced tensile strength, HIP specimens' fatigue properties were found to be superior [61].

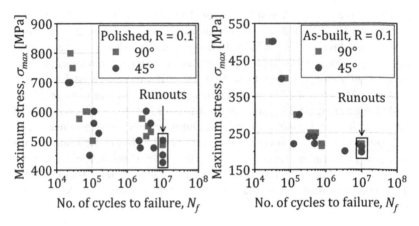

Figure 2.9 Influence of surface condition on fatigue strength of Ti-6Al-4V according to Wycisk et al. [55]

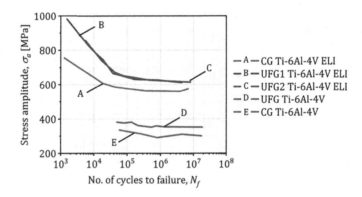

Figure 2.10 Influence of microstructure of fatigue strength of Ti-6Al-4V ELI under fully-reversed loading (stress ratio $R = -1$) according to Saitova et al. and Zherebtsov et al. [62, 63], ELI = extra low interstitial

The literature indicates that SLM Ti-6Al-4V ELI has a longer fatigue life than EBM Ti-6Al-4V ELI and commercially pure Ti. On the other hand, it is less durable than rolled Ti-6Al-4V ELI. The difference in fatigue life is mostly due to the as-built microstructural properties when the surface roughness is eliminated [64]. Due to the martensitic microstructure and the existence of flaws, the ductility was determined to be much less than that of wrought material [65]. Polishing increased the fatigue limit above 500 MPa, which is the usual value for the conventional counterpart [66]. Ultra-fine grained (UFG) microstructure has a favorable influence on Ti-6Al-4V ELI's fatigue characteristics when considering the classical *S-N* (Woehler) diagram. UFG material displays increased fatigue lifetimes due to its lower elongation-to-failure compared to the conventional grain (CG) equivalent at greater stress amplitudes, see Figure 2.10 [62, 63].

In most of the AM processes of Ti-6Al-4V, failure mechanisms depended on defects, microstructure, and other process variables [67]. EBM and SLM-produced cylindrical fatigue specimens were tested in various post-treatment conditions. In the HCF and VHCF regimes, as-built EBM and stress relieved SLM specimens have very similar fatigue properties [68]. Laser-engineered net shaping (LENS) manufacturing generated Ti-6Al-4V with greater fatigue crack growth (FCG) thresh-olds higher than EBM as-deposited Ti-6Al-4V, which showed comparable tensile strength but far higher ductility than LENS Ti-6Al-4V. Both LENS and EBM-produced Ti-6Al-4V exhibited improved low-cycle fatigue performance but inferior high-cycle fatigue performance compared to mill-annealed Ti-6Al-4V [69]. LENS of Ti-6Al-4V had contributed to shorter fatigue lives during low cycle fatigue (LCF) and HCF. Lack of ductility, as-built defects, and annealing of LENS samples had led to shorter endurance. Post-manufacturing procedures like HIP can increase the fatigue properties of the (direct laser deposited) DLD components by decreasing pore sizes [70]. Heat treatment of technical components made of Ti-6Al-4V is nec-essary to eliminate residual stresses and extend the component's life [71]. The high temperature (500 °C) of the powder bed and the relatively slow cooling rate dur-ing EBM lead to $\alpha + \beta$ microstructure. At low-stress levels, the fatigue behavior was superior in EBM specimens. This led to an approximately 20% drop in SLM specimens strength [72].

Due to overload effects and local plastic deformations, most aerospace com-ponents undergo variable amplitude loading. When the load sequence is tension-dominated, conventional cumulative fatigue-life predictions are reliable for rough surface AM materials [74]. The cyclic mechanical characteristics of porous Ti-6Al-4V investigated are within the stated range of bone mechanical properties. Curves of strain vs. cycle number showed the three-stage pattern typical of porous metals. Although the stress level was as low as $0.2 \cdot \sigma_y$, none of the configurations evalu-ated could survive 10^6 load cycles under compression [75]. A study used acoustic emission (AE) to monitor the fatigue crack growth behavior and severity in notched specimens [76, 77]. It was also shown that fatigue damage mechanisms depend on the loading domain HCF, LCF. Thus, the effect of build orientation, in particular, was not consistent with respect to the fatigue loading domain, see Figure 2.11 [73].

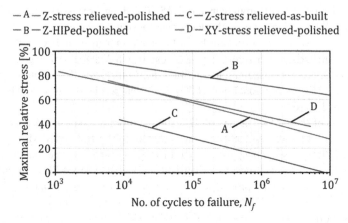

Figure 2.11 Influence of heat and surface treatment on fatigue strength of Ti-6Al-4V according to Chastand et al. [73]

HIPing heat treatment and stress relieving had no discernible effect on crack growth resistance [79] although selective laser sintered (SLS) specimens achieve reasonably high ductility and yield strength [80]. The fracture toughness was determined using an R-curve technique, which was shown to be a good way of quantifying the effect of defect distributions and microstructure [81]. Fatigue degradation processes were also associated with unique microstructural characteristics of the materials. The presence of unstable acicular martensitic α' increases tensile strength but decreases ductility [82]. The fracture toughness (K_{IC}) and fatigue crack growth rate (FCGR) da/dN characteristics of SLM specimens made from grade 5 Ti-6Al-4V powder metal showed that the fracture plane is perpendicular to the build direction [83]. In Figure 2.12, the residual stress within SLM-consolidated Ti-6Al-4V compact tension specimens is predominantly directed towards the building direction. The residual stress impact on fracture behavior is also visible during fracture toughness and FCGR tests. The unique effect of stress-reduction treatment on FCGR behavior provides additional evidence of chronic stress involvement [78, 84].

The tendency of crack propagation deviates in specimens with a microstructural α/β interface in wire + arc additive manufacturing (WAAM) Ti-6Al-4V. A fracture produced at the contact is found to propagate into the substrate, which has an equiaxed microstructure and decreased resistance to fatigue crack propagation due to the residual stresses in the original WAAM. Cracks maintained a straight

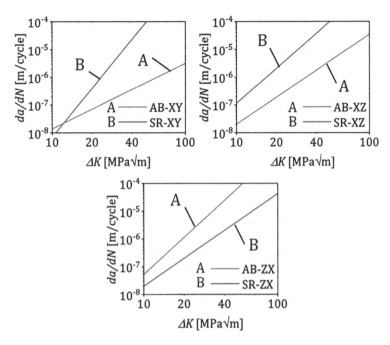

Figure 2.12 Influence of plane of loading and heat treatment on fatigue crack growth of Ti-6Al-4V according to Vrancken et al. AB: as-built; SR: stress relieved [78]

path due to symmetrical residual stress distribution [84]. In WAAM Ti-6Al-4V (see Figure 2.13), μ-CT revealed a spread of spherical pores (i.e., isolated from each other). Fatigue life and stress intensity factor (SIF) of the crack initiating defect were found to be substantially related. It was found that fractures were initiated preferentially from the defects [85]. μ-CT was used to monitor crack initiation and growth. This showed that the initiation stage represents a significant portion of life [86]. The tensile and fatigue crack growth (FCG) properties of the Ti-6Al-4V alloy at 400 °C were not significantly different from room temperature [87].

Figure 2.13 Influence of crack initiation pore on fatigue strength at tension-tension loading (load ratio $R = 0.1$) of WAAM Ti-6Al-4V according to Zhang et al. [85], WAAM = wire + arc additive manufacturing

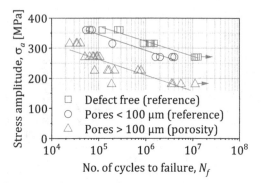

2.2   Fatigue Analysis and Modeling

Even though many fatigue life prediction models have been suggested, none of them has been widely accepted by scientists. Investigators from all over the world are devoting their time and effort to altering and expanding the current theories in order to account for all the factors that play critical roles during cyclic applications of loading [3]. Cumulative fatigue damage analysis is critical for predicting the life of components and structures that have been exposed to field load histories. Since Palmgren introduced the notion of damage accumulation around 70 years ago and Miner introduced the linear damage rule approximately 50 years ago, the management of cumulative fatigue damage has gained increased attention. Six types of theories are regarded: linear damage rules; nonlinear damage curve and two-stage linearization procedures; life curve modification methods; and crack growth concepts-based approaches [88]. A significant portion of the lifetime of high-performance materials is overshadowed by microstructural damage accumulation (e.g., extrusions and microfracture development). This necessitates statistical confirmation at the same scale by micromechanical experimentation. Effective fatigue design and prognostics are critical for the safe and sustainable operation of aircraft. The existence of structural fractures that spread over the life of the structure is a typical feature of all metallic fatigue failures. There are techniques for predicting the formation of fatigue cracks that have generally assured flight safety. However, unanticipated fatigue cracking and non-critical failures continue to occur, increasing the expense of aviation maintenance [89].

2.2.1 Deterministic and Probabilistic Approaches to Fatigue

A critical aspect of developing a new product using additive manufacturing (AM) is predicting the material's mechanical characteristics. Numerous experimental data demonstrate that AM-based products frequently exhibit significant porosity, low-density areas within their volume, and anisotropy. An article discussed the numerical modeling of a geometrically complicated structural bracket for aerospace applications that was redesigned using topological optimization techniques [90]. A technique used finite element meshes to create the volume's tessellation into a sequence of tetrahedrons whose edges form the lattice's struts. During this technique, graded porosity was defined for numerical simulation purposes. Beam finite elements are used to represent the struts to simulate the whole spectrum of porosity accurately [91]. The effect of inclusions on fatigue lifetime was calculated using a numerical model. The model is composed of multiple representative volume element (RVE) computations combined with a crystal plasticity (CP) constitutive model. RVE comprises inclusions that vary in size, shape, roughness of the surface, and elastic misfit with the matrix. The data indicate that the size of the inclusion has the greatest effect on the lifetime. This can assist in reducing the experimental effort required to determine a steel's minimum purity requirement to ensure structural integrity under cyclic loads [92]. Purdue University researchers discovered that high normal stresses function as a fracture pushing factor. In polycrystalline materials, fatigue fracture initiation is reliant on the local microstructure and deformation process, see Figure 2.14. Fatigue fractures typically begin at twin boundaries in nickel-based superalloys. Scatter in fatigue life is related to the diversity of the microstructure [93].

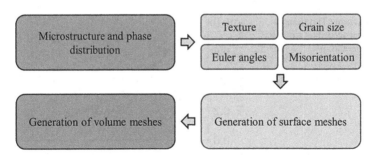

Figure 2.14 A microstructure-sensitive fatigue simulation workflow according to Yeratapally et al. [93]

   When the applied stress level on additively produced porous biomaterials did not exceed 60% of the yield stress, a computational method could predict the fatigue behavior of porous materials with varying porosities. The model proved ineffective at greater levels of stress and overestimated the fatigue life. It was discovered that these anomalies significantly decreased lifetime, particularly for specimens with lower stress levels [94]. The impact of cell shape and relative density on the fatigue behavior of titanium scaffolds manufactured using SLM and EBM methods were studied computationally using finite element analysis (FEA). All build types projected fatigue life and damage pattern were found to be in excellent agreement with experimental fatigue investigations reported in the literature. However, it should be noted that this can differ from the real fatigue characteristics of struts [95]. The load concentration on a particular structural component connected with a quarter-turn heavy-duty valve actuator, referred to as the scotch yoke, was analyzed using the Fatemi-Socie parameter. The fatigue life of the component is estimated using a full-model FEA with fillet-welded joints subjected to a constant amplitude in-phase cyclic bending torsion fatigue loading [96]. When the applied stress level did not exceed 60% of the yield stress, a computational method could predict the fatigue behavior of porous materials with varying porosities, see Figure 2.15. The model proved ineffective at greater levels of stress and overestimated the fatigue life. It was discovered that process inconsistencies significantly decreased fatigue lives, especially for those with lower stress levels [94]. A study proposed modeling fatigue damage under service loads without the need for path-equivalent range estimators or rain flow counters. Therefore, there was no accumulation of fatigue damage over time [97].

Figure 2.15 Experimental vs. simulated fatigue life of biocompatible Ti porous structures according to Hedayati et al. [94]

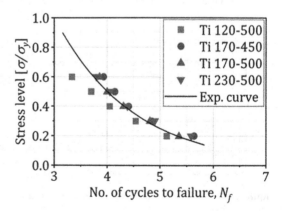

Crack propagation

In all loading situations, strain ratcheting occurs around the fracture tip, resulting in a progressive build-up of tensile strain normal to the crack development plane. Accumulated plastic strain is shown to be linked with ratcheting strain, and the results appear to be particularly positive in terms of predicting crack development rates when oxidation is eliminated [98]. A computational technique for quantitatively forecasting the development of fatigue cracks in AA2198-T8 C(T)100 specimens using one line of laser heating with a defocused laser. Residual stresses were introduced, therefore retarding the formation of a fatigue crack. The proposed modeling method could readily be applied to various procedures aimed at slowing the growth of fatigue cracking [99]. A comprehensive investigation was conducted to further verify a proposed Uni Grow crack growth model and to determine its predictive capabilities and limitations for short and long crack development behavior. The model demonstrated a strong connection with lengthy crack data sets for aluminum alloys 7010-T7, 2024-T3, 2324-T3, 7050-T7, and 7075-T6 at four different R-ratios. The model, however, did not correlate well with the short crack data set for these three materials and was unable to correctly account for short crack growth variations [100].

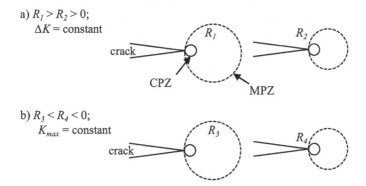

a) $R_1 > R_2 > 0$;
   $\Delta K = $ constant

b) $R_3 < R_4 < 0$;
   $K_{max} = $ constant

Figure 2.16 Illustration of plastic zone crack tip at: a) positive; b) negative stress ratios according to Kujawski [101]

A crack driving force, which does not rely on dubious crack closure statistics, is computed only utilizing the positive portion of the applied stress intensity factor (SIF) range ΔK and its highest value K_{max}, see Figure 2.16. It combines the approach for predicting the overall crack rate in terms of load ratio impacts for both long- and short-crack growth tendencies. When the predictions are compared to experimental data, a reasonable degree of consistency in observing the influence of stress ratio on monotonic (MPZ) and cyclic plastic zones (CPZ) is found [101]. The maximum rate of the fracture propagation threshold stress occurs during crack initiation in blunt notches. The peak value distributes over a limited length for acute notches. Stress greater than the initiation level but less than the peak stress level initiates crack but does not proceed to failure [102]. A novel crack driving factor is presented for the connection of load ratio effects on the development rate of long- and short-cracks in aluminum alloys. This new value, $(K + K_{max})^{0.5}$, is computed as the geometric mean of the positive component of the applied stress intensity factor (SIF) range K_{max}. The suggested parameter corresponds quite well with the R-ratio and impacts the threshold condition and fatigue crack development rate for six different types of aluminum alloys studied at low and moderate stress intensities [103]. A single crack growth law equation has been applied to the crack growth data of medium carbon steel to describe both microstructurally short and prolonged crack periods. Two crack growth rate limit curves fit the experimental results well. The upper and lower bound curves are calculated using the lowest and maximum values of the ratio $n = a/c$, where c is half the length of the surface crack, and a is half the size of the plastic zone [104]. The crack opening displacement concept is effective in describing the development of fatigue cracks up to the commencement of fracture via tearing dimple creation, see Figure 2.17. Calculating K_{th} value as a function of material, environment, and loading conditions should give a helpful, systematic engineering technique for calculating fatigue fracture development rates and determining residual lifetimes [105].

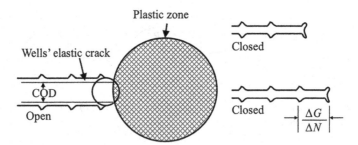

Figure 2.17 Illustration of the propagation mechanism that explains the connection between crack opening displacement (COD) and the plastic zone ahead according to Donahue et al. [105]

At zero load ratio, a fracture propagation under zero-to-tension stress can be closed partially or entirely. Local compressive stress maxima can surpass the material's yield stress. Fractographic evidence indicates crack closure can have an effect on the morphology of the striation pattern on the fracture faces [106]. The range of fracture tip critical stress intensity factors K_c, is the governing variable when assessing crack extension rates. A hypothesis incorporates the load ratio R, and the instability that occurs when the stress intensity factor approaches the material's fracture toughness. The theory and substantial experimental data exhibit an excellent connection [107]. Compression simulations of porous Ti filled with polymer are used to compare the mechanical behavior of purely porous and composite materials. The results suggest a high degree of congruence between simulations and experiments [108]. At room temperature and under different flexion bending cycling circumstances, the fatigue damage build-up for block program loading step-down sequences with two-step and three-step through amplitude change showed an increased fatigue strength per block program-step to compensate for the lower consumed fatigue loading. While in the case of a block program step, residual fatigue strength is increased due to a greater number of consumed fatigue cycles, see Figure 2.18. The tendency underestimates fatigue life and cumulative damage exceeds unity [109].

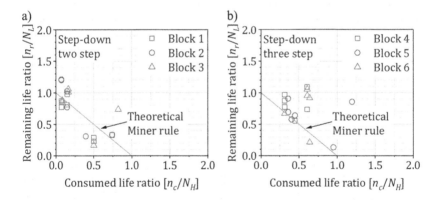

Figure 2.18 Fatigue strength increase compared using experimental and fatigue lifetime predictions in aluminum 7050-T7451 alloy: a) two-step; b) three-step variable amplitude loading blocks according to Carvalho et al. [109]

A model for multiaxial fatigue life prediction is developed using a critical plane model and the equivalent initial flaw size (EIFS) technique. The suggested model is capable of automatically adapting to diverse materials with varying local failure mechanisms. Under both proportional and nonproportional loading, a reasonable agreement between model predictions and experimental results is obtained. The critical plane is theoretically defined by the maximum normal stress plane and the ratio of the coefficients of modes II and I stress intensity factors related to the crack growth rate [110]. Crack initiation was numerically simulated using Abaqus and a plug-in developed specifically for addressing microcrack nucleation and coalescence. Numerous slip bands were utilized in each grain to simultaneously produce multiple fractures [111].

Crack initiation along slip planes is largely dependent on the local stress in polycrystalline materials. The Tanaka-Mura model establishes a quantitative relationship between the slip line length, the resolved shear stress along this line, and the number of cycles required to initiate a fracture [112]. An article described a numerical simulation of the formation of microcracks using the Tanaka-Mura model. It considered numerous slip bands in which microcracks can form in each grain. Three enhancements are made to this model, including macrocrack coalescence and segmentation. Additionally, HCF testing was done, which demonstrated a strong agreement between the suggested model and the experimental findings [113]. Fracture and crack propagation in polycrystalline materials was modeled using the multiscale cohesive technique. It can be used to effectively describe heterogeneous material properties at a small scale by accounting for the effect of inhomogeneities. Numerical simulations of crack propagation over a range of crack strengths demonstrated the capability to capture the shift from intergranular to transgranular fractures [114]. Short crack behavior of Ti-6Al-4V is produced by incorporating the lowered threshold caused by unsaturated closure into long fracture propagation. By estimating the fatigue life of a notched specimen, the model was validated. When propagation was significantly less than the initiation, and when virtually the fatigue life was mostly initiation, an accurate forecast was achieved [115].

To simulate the inhomogeneous stress distribution and its effect on multi-crack initiation behavior in martensitic steel, a computer model based on a simplified three-dimensional model of the microstructure is defined. In this simulation, the Tanaka-Mura relation for crack initiation is utilized to determine the number of cycles required for a crack to start in each grain [117]. The virtual crack extension method is used to simulate the propagation of cracks within the context of FEA, see Figure 2.19. Due to its theoretical superiority in dealing with mixed-mode short crack propagation close to the loading border, it more precisely forecasts the crack route and functional relationship $K = f(a)$. However, the strain energy density (SED) criterion's reliability might be increased by incorporating appropriate constraint parameters that allow for a more precise calculation of the stress intensity

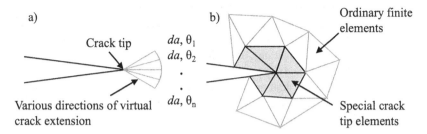

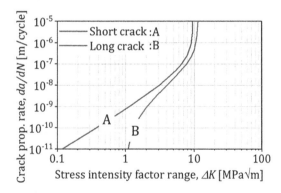

Figure 2.19 Illustration of crack extension scenario based on the virtual crack extension principle according to Fajdiga et al. [116]: a) various possible directions; b) enriched zone around the crack tip

components [116]. Fatigue crack propagation threshold was identified as a function of crack length. The material threshold is calculated using the plain fatigue limit and the location of the most effective microstructural barrier to fracture propagation. Excellent agreement was reported for eight distinct material systems [118].

Figure 2.20 Distinction between short and long crack propagation for 7050-T7451 according to Jones et al. [119]

A new method for correlating load ratio effects on the development rate of long- and physically short cracks in aluminum alloys is predicated on the premise that the perceived efficacy of the crack closure phenomenon can be different for long fractures than for physically short cracks. The improved partial crack closure model is used to describe the growth behavior of long fractures, whereas the traditional Elber's closure model describes the physical short cracks' growth behavior [120]. From the tensile characteristics of industrial metals, a novel expression for fatigue crack propagation can be anticipated. The proposed expression is consistent with fatigue crack propagation results for steels with a rate less than 10^{-3} mm/cycle and accurately describes the influence of stress ratio [121]. At a variety of R ratios, it was discovered that crack growth in twenty-two different materials follows a version of the Hartman-Schijve equation. Additionally, it was proven that this formulation holds true for both long and short fractures in 7050-T7451 but that a different value of K_{th} was necessary for short cracks vs. long cracks, see Figure 2.20. The fatigue crack rate da/dN equation established of a vast quantity of crack data can result in a simple method for forecasting the lifetime of fatigue cracks [119].

Microstructure-sensitive simulations
Microstructure-aware design and prognosis procedures comprise determining the effect of microstructure morphology on the fatigue response scatter. Recent applications of these techniques to superalloys based on nickel or gear steels and titanium alloys are appreciably found in the literature for conventional and additively manufactured alloys [122].

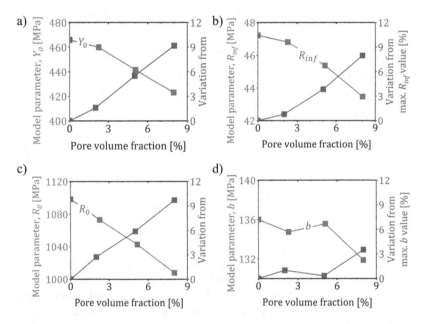

Figure 2.21 Influence of volume fraction on mechanical properties according to R. G. Prasad et al. [123]

The anisotropic mechanical behavior of SLM manufactured 316L in the presence of pores was investigated using a micromechanical model. The form of the pores influences not only the yielding and hardening behavior but also the evolution of porosity, according to Figure 2.21. Y_o, R_o, R_{inf}, and b are the model parameters. When loaded orthogonally, microstructures with elongated pores exhibit considerable anisotropy with increasing pore volume fractions and an earlier beginning of damage [123]. A three-dimensional crystal plasticity model was developed for a duplex Ti-6Al-4V alloy to analyze grain size plastic deformation. To calibrate and validate the model, extensive experimental deformation and damage modes were generated. Most single slips were seen at the projected LCF strain. This property was enhanced by applying a softening evolution algorithm to the threshold stress of each active slip mechanism [124]. Another strategy is to raise the microstructural short crack (MSC) propagation threshold by lowering barrier spacing, encouraging stage I propagation extension, or increasing barrier strength. In the absence of robust and accurate damage state detection techniques capable of sensing and separating tiny fatigue fractures, this type of system can provide information about the predicted distribution of cracks propagating to detectable lengths [122]. A model incorporated length scale effects caused by dislocation interactions with various microstructural characteristics. It is calibrated using polycrystalline finite element simulations in order to suit the observed macroscopic responses (overall stress-strain behavior) of a duplex heat-treated alloy exposed to a complicated loading history, see Figure 2.22 [125]. It is important to take into account the unique slip geometries of both phases, length scale-dependent and anisotropic slip strengths, non-planar prismatic dislocation core structure, and crystallographic texture [126].

The endurance and HCF lifetime of multiphase steel components are mostly determined by the phase of microcrack formation and early propagation during fatigue. The damage sliding mechanisms in S355J2 was used with the presence of inclusions/voids to improve the prediction of fatigue fracture development in this material system [127]. Using the finite element method (FEM) and the computation of the material's non-uniform stress distribution, a micro-model including the material's microstructure was produced. The number of cycles required for crack initiation was estimated using the microstructural model's stress distribution and the physically-based Tanaka-Mura (TM) model. The lengthy crack development was investigated using Paris law and found to be in good accord with published experimental data [128]. A revision of the 1981 micromechanical model for the start of a fatigue fracture was proposed by K. Tanaka and A. Mura [129]. The main limitation of the model relies on the empirical observations which were previously established by Weibull [130]. The TM model captures the influence of loading on the fatigue crack development rate during the initiation phase due to the incorporation of the

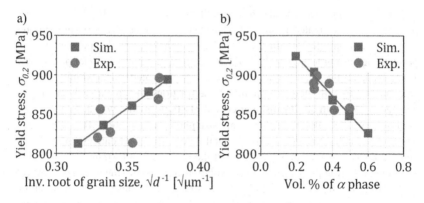

Figure 2.22 Influence of microstructure on mechanical properties according to Zhang et al. [125]: a) grain size; b) fraction of α phase

TM equation in the finite element (FE) method-based modeling. While the elastic material model captured the increase in crack growth rate caused by overload, the expected retardation could not be observed due to a lack of material plasticity [131]. Finding the constants C and m in the Paris law numerically was achieved by a two-scale model was successful. The stress intensity factors were computed using the macro model based on traditional LEFM for six distinct structural splits. In the second stage, a microstructural model enhanced with the TM equation was used to calculate the rate of fatigue fracture development at the front of such cracks [132].

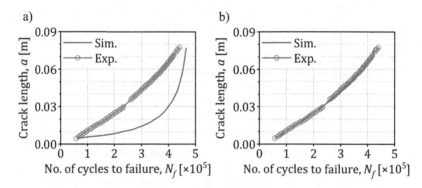

Figure 2.23 Validation of crack growth simulation under the influence of residual stresses: a) without residual stresses; b) accounting for residual stresses according to Bozic et al. [133]

A multiscale numerical method was presented for predicting the HCF strength. The objective is to link macroscopic fatigue characteristics with microstructural traits and their associated micro deformation processes. The methodology is comprised of a set of numerical and mechanism-based analytical models. The most plastically deformed grain was selected in each representative volume element (RVE) and its average plastic strain value was collected for further investigation [134]. It is shown that in the region of a stiffener, high tensile residual stresses raise, the simulated crack development rate was faster, which is consistent with experimental data, see Figure 2.23 [133]. The energy-based growth formulation and associated simulation technique were proposed to employ a new basis-function approach to create a crack extension expression. The method mitigates the influence of numerical noise on crack development forecasts, which would otherwise result in numerically unstable simulation results [135]. Dislocation interactions in bands of intense cyclic slip have defined persistent slip bands (PSB) dislocation movement patterns in the direction of the active Burgers vector that relaxes local tensile and compressive stresses. As a result, distinctive surface relief in the form of extrusions and intrusions is formed. At low temperatures, intrusions indicate crack-like flaws, and fatigue fractures begin at the tip of intrusions [136]. To mimic the localization of deformation and the propagation of microstructurally small fatigue fractures in crystalline materials, a mesoscale model based on FE was created. The inelastic hysteresis energy is used to forecast the onset and spread of fatigue cracks. It accurately forecasts the start of cracks on slip bands and at inclusions in LCF and HCF [137].

Figure 2.24 Cycles to crack nucleation in high strength steel: Tanaka-Mura (T-M); Miller Hobson (M-H) according to Mikkola et al. [138]

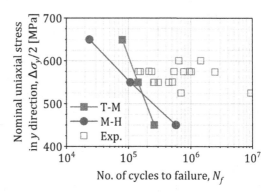

The initiation of a microcrack near a defect was highly dependent on the orientation of the surrounding grains, and only a small percentage of faults of a given size resulted in fatigue cracking. This was not the case even when grains near the defect had low Schmid factors. This resulted in a conservative estimation of cycles to crack initiation, as shown in Figure 2.24, where σ_y is the nominal uniaxial stress in the y-direction [138]. Dual-phase (DP) steels are characterized by ferrite and martensite constituents as well as cleavage fracture. The researchers investigated the interplay of failure modes in DP steels using two distinct damage mechanics techniques and a representative volume element (RVE) based on the extended finite element method (XFEM) method. This explained the relationship between ductile damage in the ferrites under multiaxial stress [139]. An article discussed an approach for creating artificial microstructure models to predict the strain-hardening behavior of DP steel. The flow curves were consistent; however, local deformation was not similar since grains' deformation behavior varied according to their orientation. This can result in microcracks such as situations encountered in sheet metal forming uniaxial tension, plane-strain tension, and equi-biaxial pressure. The damage is initiated as a result of the extreme plastic strain localization at the interface between ferrite and martensite induced by these two phases' incompatible deformation behavior [140]. The fatigue reliability of a lap shear riveted joint was evaluated using detail fatigue rating (DFR) and FE simulations using a rivet element. The traditional DFR technique requires global FE analysis of the whole structure to discover joint loads and stress concentration factors, compromising computing time and accuracy. The introduced DFR-Rivet element method reduces calculation time by at least 95% for analyzing fatigue reliability of real structures [141]. A mechanistic technique for modeling the formation of microstructurally sensitive (tortuosity and propagation rate) fatigue cracks in ductile metal simulates the fracture. The model performs crystal plasticity finite element (CPFE) simulations using the extended XFEM by incorporating a grain boundary. It is demonstrated that the crystallographic model accurately reproduces empirically observed behavior, including fracture deflection and retardation at grain borders [142]. XFEM was used to investigate the mechanical determinants of fracture path tortuosity and propagation rate. Fracture propagation was discovered to be governed by the stored energy at the crack tip and the direction of the crack by anisotropic crystallographic slip at the crack tip. Local slip dominates the formation of very small cracks, but as they become longer, the crack tip state takes precedence [143].

Probabilistic analysis

Canteli's fatigue model was generalized to numerous fatigue damage features in a probabilistic model, including Smith-Watson-Topper, Walker-like strain, and energy-based parameters under multiaxial loading circumstances [145]. When compared to experimental data for metallic materials, Canteli's fatigue model demonstrated good adaptability from LCF to HCF fatigue regimes. The probabilistic Stüssi and Sendeckyj models for composite materials demonstrate to be successful across all fatigue areas, see Figure 2.25 [144]. The experimental fatigue data from structural steel plates with a high strength low alloy content were evaluated statistically using linearized boundaries and the Castillo and Fernández-Canteli model. The use of bonded and bolted connections is proven to be feasible. The fatigue strength was evaluated utilizing an acrylic structural adhesive for metal bonding [146]. Models of multiaxial stress damage usually lack probabilistic interpretation due to their deterministic nature. The predominant defect's random orientation evidences the possibility of failure being initiated as a function of the predominant defect's presence. The probability of failure was calculated by considering the damage gradient [147].

An article proposed a novel approach for calculating the EIFS distribution. The Kitagawa–Takahashi diagram serves as the foundation for the suggested technique. Unlike the frequently utilized back-extrapolation approach, the suggested probabilistic methodology utilized just the fatigue limit and fatigue crack threshold stress intensity component for calculating EIFS, see Figure 2.26. It does not address the inverse crack development problem, which greatly simplifies the generation of probabilistic EIFS data [148].

Figure 2.25 Probabilistic fatigue lifetime prediction of laminate DD16 according to Barbosa et al. [144]

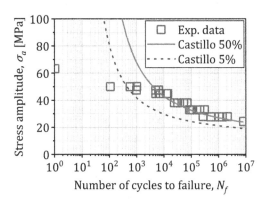

A statistical method should be used when examining the effect of material microstructure (grain orientations) on structural element fatigue behavior. The effect of various grain orientations produced by Voronoi tessellation on the crack initiation period was investigated through numerical modeling [149]. For cracked structures, a strategy was utilized to construct a universal size effect model and a probabilistic

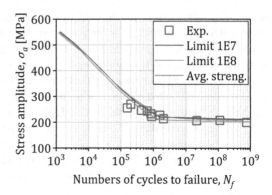

Figure 2.26 Influence of equivalent initial flaw size on the statistical fatigue strength according to Liu and Mahadevan [148]

fracture analysis. It demonstrated statistically yield values for energy release rates, their first and second-order derivatives, that are expectedly decreasing yet acceptably accurate [150]. Crack growth was modeled in notched components subjected to multiaxial strain. The method for crack development was developed in competition with the propagation of a dominant crack via crack connections. The results were compared to experimental data obtained using notched specimens of pure copper, carbon steel, and two different kinds of titanium alloy [151].

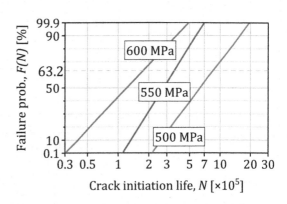

Figure 2.27 Dependence of Weibull line median values on stress level according to Glodez et al. [152]

Researchers at the University of Maribor have devised a novel method for evaluating microcracks in high-strength steel, see Figure 2.27. They discovered that when they subjected their high-strength steel structures to compressive loading conditions, the edges developed a micro-crack. This is because each component/grain has edges with unique contact characteristics [152]. A probabilistic model based on Monte Carlo theory is used to predict the fatigue behavior of cast aluminum alloys. It was aimed at the determination of the effect of porosity (defect size distribution and spatial density) on fatigue strength under uniaxial fatigue conditions. Although the scatter is underestimated, it is equivalent to experimental data. This is because the proposed model assumes that the only source of dispersion is porosity [153].

2.2.2 Process and Microstructural Simulation

To assess the basics of grain structure evolution during metal additive manufacturing, a three-dimensional numerical model was constructed. The grain structure is predicted using cellular automata and finite difference techniques in response to a transient temperature field generated during the additive manufacturing process. The results indicate that SLM, in combination with the competitive character of grain growth, promotes the production of coarse columnar grains [154]. The effect of temperature gradients on the evolution of part microstructures was simulated through the evolution of the microstructure of Ti-6Al-4V powder during the EBM process using a phase-field model. Figure 2.28 shows that the dendritic arm width will decrease with the increasing scanning speed due to higher cooling rates [155].

Figure 2.28 Influence of scan speed on dendritic width according to the phase-field method applied by Sahoo and Chou [155]

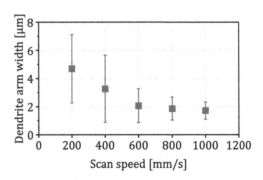

Thermally generated residual strains and residual distortions in AM components are two key impediments to the general adoption of AM technology [156]. A model could correctly replicate and preserve the whole temperature history of a two-dimensional component (or a thin-walled 3D part) produced in a powder bed fusion (PBF) AM method. This was accomplished using an adaptive meshing technique, which significantly reduces the computing time and memory requirements for the simulation. It enables the user to adjust the process parameters in such a way that the user is assured of the optimal cooling rates [157]. A study described an attempt to model and forecast the performance of elastic AM structures, see Figure 2.29. The relationship between specific heat and thermal conductivity and temperature has been substantially simplified. Because all the findings are numerical, an experimental test program is necessary to fine-tune the FE model. The approach can be used to provide a valuable basis of information regarding the deformations and residual stresses of various materials (aluminum, titanium alloy, steel, etc.) [158].

Figure 2.29 Dependence of elastic modulus on temperature according to Conti et al. [158]

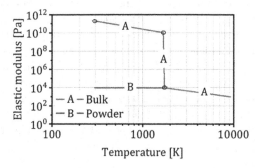

In a COMSOL multiphysics environment, a numerical model was built to account for the temperature-dependent material characteristics of Ti-6Al-4V during direct metal laser sintering (DMLS). The results demonstrate the behavior of a single layer's melt pool size, temperature history, and change in residual stresses [159]. The influence of stress relaxation at elevated temperatures on laser direct energy deposition (DED) processes is experimentally confirmed using Ti-6Al-4V samples with varying inter-layer dwell durations. The model predicts that whereas Inconel exhibits growing distortion as dwell durations decrease, Ti-6Al-4V exhibits the opposite trend, with distortion significantly reducing as dwell time decreases [160].

A 3D transient fully coupled thermomechanical model was developed to investigate the distortion and residual stress in Ti-6Al-4V build plates produced by EBM. Simulated and experimental investigations of single layer, six-layer, and eleven-layer build plates were conducted to confirm the model's correctness and to instill confidence in the use of simulations for process optimization. The results of the simulations indicate that preheating efficiently reduces the final distortion and residual stress [161]. A study discussed the creation of a methodology for modeling particle-based AM processes using thermal physics. It has shown that this simulation framework could capture the spatial and temporal variations in heat and mechanical stress distributions within an additively produced item [162].

Figure 2.30 Remelt volume fraction during remelting at various scanning speeds according to Vastola et al. [163]

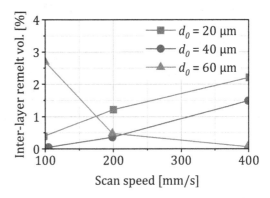

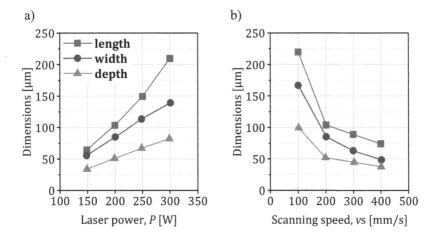

Figure 2.31 Influence of process parameters on melt pool width dimensions according to Li and Gu [164]

Research aimed to ascertain the melt pool shape and temperature distribution inside a powder bed. A model was used to compare the Ti-6Al-4V, stainless steel 316L, and 7075 aluminum powders on these properties. Steel powder beds require more beam power to produce a constant melt pool than titanium powder beds. To maintain a complete melt pool in aluminum, the beam power must be raised while the scan speed is reduced [165]. Researchers conducted an explicit investigation of remelting in SLM and EBM using computer simulations. They discovered that there is an optimum beam shape for remelting at a constant beam cross-section. When the scan speed is altered, the geometry of the beam varies, and interlayer removal is addressed explicitly in this scenario, see Figure 2.30. The results indicate that varied beam cross-sections have a significant influence on the remelt volume and, ultimately, on the component quality [163]. When the laser power was increased from 150 to 300 W, the molten pool's cooling rate increased slightly from 2.13 to 2.97 °C/s. The requisite molten pool width (111.40 μm) and depth (67.50 μm) for a successful SLM operation were achieved with a combination of power P of 250 W and scanning speed vs of 200 mm/s, according to Figure 2.31 [164].

Training and Testing Data

3

3.1 Aluminum and Titanium Lightweight Alloys

3.1.1 Aluminum (Al) Alloys

Why did Al alloys become attractive?
Aluminum is a post-transition metal that is part of the periodic table's third era. Its crystal structure is face-centered cubic (FCC), with a density of 2.70 g/cm^3 and an elastic modulus of 70 GPa. Sir Humphry Davy discovered the presence of aluminum in the alum salt that bears its name [166]. Improved static strength and resistance to fatigue, crack growth, and fracture toughness have been the driving forces for the development of Al-alloys in aerospace frameworks. Aluminum's FCC structure and a large number of slip systems aid in plastic straining to accommodate local damage [167].

Classification of Al alloys
Al-alloys can be differentiated into cast and wrought alloys. Numerous cast and wrought alloys exhibit phase solubility, solution, quenching, precipitation, and age hardening responses to thermal treatments. These alloys are reputably heat treatable since distinct phases, and hence distinct characteristics can be generated by manipulating the solid-state precipitation [2]. An alloy nomenclature system can be used to classify both wrought and cast alloys [168]. Aluminum and its alloys can be cast or molded in almost any shape or configuration. Porosity and other apparent flaws are frequent in solid-solution-processed components. Over

---

Some content in this chapter is partially based on publications [183] and [191]

© The Author(s), under exclusive license to Springer Fachmedien Wiesbaden GmbH, part of Springer Nature 2022
M. Mamduh Mustafa Awd, *Machine Learning Algorithm for Fatigue Fields in Additive Manufacturing*, Werkstofftechnische Berichte | Reports of Materials Science and Engineering, https:doi.org/10.1007/978-3-658-40237-2_3

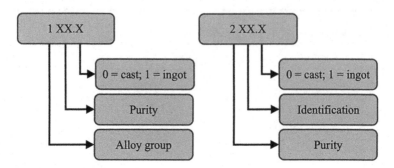

Figure 3.1 A coding system designation for cast aluminum alloys [2]

the years, several ways have been developed to manage or, in some circumstances, remove porosity. Degassing, filtration, and improved handling of the molten alloy are just a few examples [169, 170]. A coding system for cast Al alloys is shown in Figure 3.1.

Wrought alloys: Extrusions, foils, forgings, plates, sheets, stampings, and wires are examples of wrought items. The non-heat-treatable Al sheets are the most popular type, and it is widely employed in the building, packaging, and transportation industries. The most common feedstock for metalworking is an alloy ingot made by direct cold casting [169]. The thickness of wrought aluminum sheet products is normally between 0.15 and 6.30 mm. Plates have a thickness of more than 6.30 mm. Nomenclature can be found in Figure 3.2 [171].

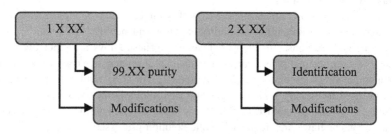

Figure 3.2 A coding system designation for wrought aluminum alloys [2]

Controlled heat conditions, for example, can result in rapid melting and solidification, as in welding. Cold working methods such as shot peening and comparable operations can significantly alter an alloy's mechanical characteristics

[172]. Silicon solid particles that do not dissolve in aluminum create a biphasic mixture consisting mostly of pure silicon granules scattered inside an aluminum matrix (see Figure 3.3). Silicon is employed in welding procedures in the 4XXX series due to the existence of a low-temperature eutectic transition [170]. This character qualifies these alloys for AM processes.

Heat treatable aluminum alloys have an international designation code that is appended to the alloy designation number. To harden an alloy, microstructurally, stable or metastable precipitates must be developed by the heat treatments, which comprise a solution-phase strengthening, quenching, and aging. The role of relevant lattice defects in solid-state precipitation is essential in the hardening process. Most metastable structures in heat-treatable aluminum alloys are the β' precipitates in the 6063 alloy (Al-Mg-Si group, Figure 3.4). Mg2Si has a preferentially oriented precipitation orientation along with the <100> aluminum matrix with β'. This crystallographic direction turns out to be more energetically favored than the others [2].

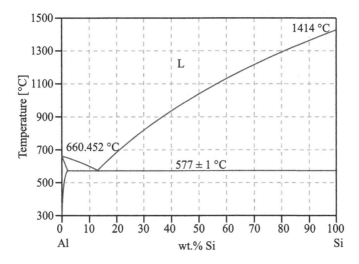

Figure 3.3 Al-Si phase diagram [170]

Figure 3.4 Schematic
illustration of strain in the
precipitate matrix coherent
interface of β' in an
Al-Mg-Si alloy [2]

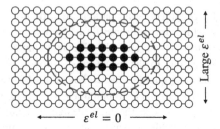

The interfaces with the aluminum matrix will be broader because the aluminum matrix has a smaller elastic strain ε^{el} energy term, see Figure 3.4. Precipitation is also susceptible to elastic anisotropy. The rate of precipitation and dislocation interaction is determined by the microstructure of the resultant structure [173].

3.1.2   Titanium (Ti) Alloys

From a metallurgical point of view
The periodic table of elements glides titanium (Ti) in the transition metal zone with an atomic number of 22. The family of allotropic alloys based on the titanium element consists of α Ti (stable up to 882 °C). This is marked as β transus temperature at which β titanium begins to be stable as temperature increases up to the liquidus line at 1670 °C [2]. The lattice structures for α and β titanium are hexagonally close-packed (HCP) and body-centered cubic (BCC) respectively see Figure 3.5. The immediate benefits of such structures are the exceptional corrosion resistance and the high specific strength [174]. The aircraft industry was the primary element behind the industrialization of titanium alloys. Ti is found mostly as oxides in the earth's crust.

The crystallographic transformation at 882 °C is common between pure and alloyed titanium with respective relevant variations in the boundary conditions depending on the metallurgical boundary conditions. The existence of either phase can be promoted and stabilized by certain alloying elements. The coexistence of both phases provides Ti alloys with a preferential set of properties that can be tailored according to the application at hand [2]. The main attributes of these are summed as follows: 1/ α Ti: the HCP lattice structure is strong with limitations on toughness and formability. By applying the von Mises criterion [176, 177], it can be deduced that at least five independent slip systems have

Figure 3.5 Lattice structures of the main allotropic elements of Ti alloys: α and β titanium [175]

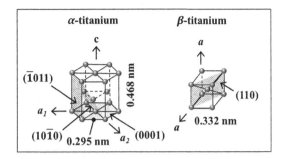

to be activated for homogenous plastic deformation to occur. 2/ β Ti: the BCC polymorph with its 12 slip systems is easier to deform but limits other properties such as creep resistance. The solid solution transforms during cooling from β to α depending on the cooling rate and composition according to two main mechanistic processes. The lath or packed martensite will experience shearing.

Classification of Ti alloys
Two main families of Ti alloys are identified, which are the traditional and modern Ti alloys [2]. In Figure 3.6, the resulting properties of traditional Ti alloys will rely on their thermomechanical history. In the modern Ti alloys, several lattice systems and metallurgical conditions are considered to achieve the desired output, such as in mainly aerospace applied TiAl-based alloys. The main contribution to the desired properties happens from the intermetallic phases.

The BCC character of traditional Ti alloys widens the application window of Ti alloys. The $\alpha + \beta$ alloys are also known as aeronautic alloys, and they turned out to be significantly more capable than pure α Ti alloys. Toughness and creep resistance are significantly improved with the introduction of β stabilizers up to the introduction of the grade 5 (IMI318) Ti-6Al-4V in 1958, which has the most attractive balanced mechanical properties [174, 175]. In this alloy, the β phase content is equal to 15%, which can be reached by β stabilizers that lower β transus. The solution and aging treatments take advantage of these metallurgical aspects to improve and tailor properties further. When alloying elements are used, the α and β polymorph stability range can be adjusted. According to the polymorph, the alloying elements can classify into α stabilizers, β stabilizers, and neutral elements. The other two sub-groups include isomorphous and eutectoid types of β stabilizers [178]. The binary Ti-Al phase diagram has a variety of intermetallic phases. Ti3Al and TiAl are of technological significance, as sophisticated TiAl alloys are based on these ordered phases, see Figure 3.7. The

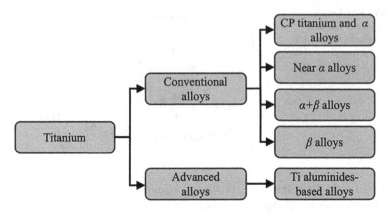

Figure 3.6 Composition-based classification of Ti alloys based on the dominant phases [2]

service temperature of ($\alpha + \beta$) alloys is limited to 440 °C, at which point creep deformation rates become unacceptably high [175].

3.2   Additive Processing and Functional Grading

3.2.1   Principles of Additive Manufacturing

How did additive manufacturing start?
The phrase "additive manufacturing" (AM) combines the concepts "rapid proto-typing" and "3D printing." Rapid prototyping [179] is a broad phrase that refers to a methodology for quickly developing a system or part model before commercialization. The term "prototyping" comes from the word "proto," which means a "quickly produced basic model" [180]. Figure 3.8 lists the basic steps from a prototype based on stereolithography standard tessellation language (STL) to the finished product.

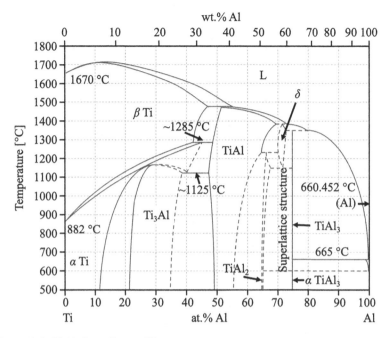

Figure 3.7 Ti-Al phase diagram [2]

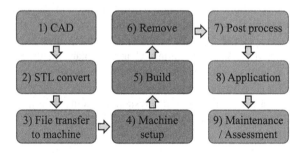

Figure 3.8 Fundamental stages of a rapid prototyping/additive manufacturing development chain [179]

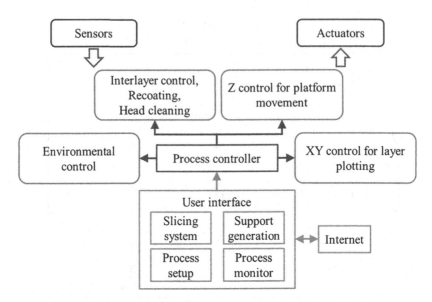

Figure 3.9 Architecture of additive manufacturing technologies implemented in modern machines [180]

Modern computer technology is included inside the AM machines as well as employed as a supporting technology. Precision placement of equipment such as a computer numerical control (CNC) machining center, high-end photocopier, or laser printer is required for AM technology. Building files are created on a separate machine while AM is operating. Complementary integration [181]: the computer is a central component that serves to coordinate different processes. The computer's goal is to communicate with other parts of the system, process data, and send data from one part of the system to another [180]. In Figure 3.9, the technologies listed are used to create an AM machine.

3.2.2 Functional Grading and Local Heat Treatments

Platform heating (PH)
Modified SLM 250 HL and SLM 500 HL systems for AlSi10Mg and Ti-6Al-4V alloys, respectively, with an external laser source capable of generating a

maximum laser power of 1000 W were used to manufacture cylindrical micro-computed tomography (μ-CT), tensile and fatigue specimen. An inert argon environment was used to keep the melt tracks from oxidizing during powder melting. High-quality components with a relative density of more than 99.50% were produced using an improved scanning technique and settings. Two contour scans are followed by bidirectional scanning of the core in this two-step scanning method. The scanning vectors are rotated by 90° layer by layer. The final geometry for mechanical testing methods was attained by further machining the produced specimens. All specimens were subjected to platform heating at a temperature of 200 °C. The parameters for one standard exposure are shown in Table 3.1. One batch of AlSi10Mg and one batch of Ti-6Al-4V were made using the scanning settings listed in Table 3.1 where laser power is (P), scanning speed is (vs), spot size is (D), hatching distance is (h_t), and layer thickness is (t). All the specimens in this research are built at a 90° angle, with the cylinders' axes perpendicular to the platform.

Table 3.1 Applied laser scanning parameters for building AlSi10Mg and Ti-6Al-4V alloys specimens in a single exposure treatment

Parameter	Power	Scanning speed	Spot size	Hatching distance	Layer thickness
	P (W)	vs (mm/s)	D (mm)	h_t (mm)	t (mm)
AlSi10Mg					
	350	1200	0.083	0.190	0.050
Ti-6Al-4V					
	240	1200	0.082	0.105	0.060

The platform heating refers to thermal energy input to the melt pool from the construction platform, which is carried upwards through the specimens. As a result, the cooling rate is slowed, and the resultant microstructure's thermal history is altered. Within the SLM process, the method allows for a degree of customization of microstructure and characteristics.

In situ heat treatments (investigations at Fraunhofer IAPT)
For Ti-6Al-4V, the impact of in situ thermal treatments, meaning simply remelting, is investigated using the experiment design in Table 3.2. The typical first exposure parameters are maintained throughout the experiment. Still, the impact of the second exposure is separated by varying the scanning variables. For the

first and second exposures, the identical scanning method was used. The second exposures begin immediately after the initial exposure for each specimen. Table 3.2 shows the applied sixteen parameter sets.

Table 3.2 Design of the remelting experiment of Ti-6Al-4V where Ev is the energy density

Parameter	Power	Scanning speed	Spot size	Hatching distance	Layer thickness
	P (W)	vs (mm/s)	D (mm)	h_t (mm)	t (mm)
Ti-6Al-4V (PH = 200 °C)					
Standard	240	1200	0.082	0.105	0.060
Second exposure					
$Ev1$	200	1000	0.082	0.105	0.060
$Ev2$	160	800	0.082	0.105	0.060
$Ev3$	120	600	0.082	0.105	0.060
$Ev4$	80	400	0.082	0.105	0.060
$P1$	80	1200	0.082	0.105	0.060
$P2$	160	1200	0.082	0.105	0.060
$P3$	320	1200	0.082	0.105	0.060
$P4$	400	1200	0.082	0.105	0.060
$vs1$	240	3600	0.082	0.105	0.060
$vs2$	240	1800	0.082	0.105	0.060
$vs3$	240	900	0.082	0.105	0.060
$vs4$	240	720	0.082	0.105	0.060
$D1$	240	1200	0.116	0.105	0.060
$D2$	240	1200	0.142	0.105	0.060
$D3$	240	1200	0.164	0.105	0.060
$D4$	240	1200	0.183	0.105	0.060

A wide-field thermal camera from EQUUS (81k SM/M) was used to track the thermal history and heat build-up of the residual porosity and its shape during the remelting process. Figure 3.10a depicts an illustration of the camera's experimental setup concerning the construction platform and specimen placement. The temperature calibration functions were set by placing a powder and solid component sample in an oven that measured temperature using thermocouples and surface radiation using the thermal camera. Figure 3.10b depicts the correlation between these numbers. Multiple specimens can be tracked simultaneously, revealing variations across applied remelting parameter settings, thanks to the setup's ability to monitor a field of view of 100×100 mm^2 with 320×256 pixels and 300 frames per second (fps). Because the diameter of the melt pool is almost half the size of one pixel, observed temperature values could not be linked to melting pool temperatures in this situation.

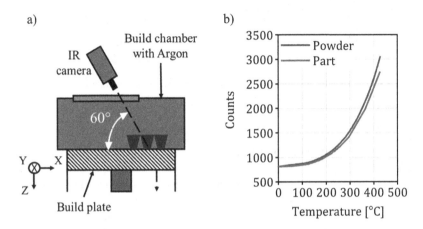

Figure 3.10 Concept of measurements of the thermal camera: a) schematic of the setup; b) graphical representation of the quantification method

Three critical factors to investigate during the examination of the remelting method's influence are the mean specimen surface temperature before recoating, the mean sample surface temperature prior to scanning, and the solid-to-solid cooling rate from 400 to 300 °C. The temperature history of a region of interest (ROI) with an estimated 700×700 μm^2 is illustrated in one layer of the remelting process *P2* Figure 3.11b. The recoater stroke causes an abrupt drop in temperature, after which heat begins to accumulate again as the temperature reaches the absolute maximum. The temperature decays exponentially in the absence of the laser until the second exposure is applied, resulting in a significant temperature rise. After the second exposure, the temperature decays exponentially again until a far slower cooling rate is attained. The mean surface temperature of a single layer within the applied remelting strategy is shown in Figure 3.11b.

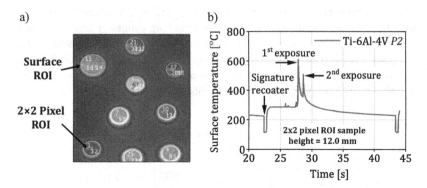

Figure 3.11 a) Real-time in situ measurements of the current layer (snapshot of measured region of interest, ROI); b) mean surface temperature of a single layer within the applied remelting strategy [183]

The thermal camera recorded the mean surface temperature of all Ti-6Al-4V combinations in two stages over the specimens' construction height. Before recoating, there is the first phase, and then there is a second phase, which is before scanning the initial exposure. The first specimens for which the thermal data are shown in Figure 3.12 are 12 mm tall and 5 mm in diameter. We can observe that heat will collect over the building height in any scenario by comparing the two stages before recoating and scanning. In the parameter set *P2*, the mean surface temperature can grow by more than 20 °C along with the height. At both measurement phases, this parameter set gathered the highest thermal energy. Single (*Sin.* in Figure 3.12) exposure heat readings are given as a comparison.

However, when comparing the starting and end temperatures, the variation in mean surface temperature for the single exposure specimen is among the smallest. In the second exposure, the parameter settings with the highest energy density accumulated the most heat. As shown in Figure 3.13a, mean surface temperature measurements of the mechanical testing specimens cross-section as an ROI were also taken and compared before recoating and scanning.

The single exposure specimen has the lowest temperature in both figures, either before recoating or before scanning, with batch *D1* having the highest due to excessive heat across layers that build up due to lower-than-normal scanning speed.

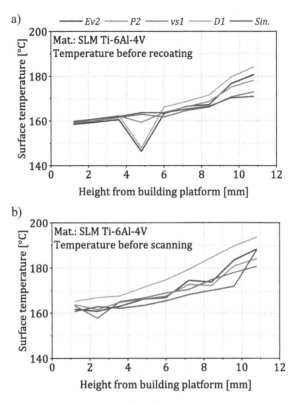

Figure 3.12 Mean surface temperature along build height in the porosity testing specimen for the microcomputed tomography specimens: a) before recoating; b) before scanning

In general, up to a height of 20 mm, all batches exhibit a build-up of heat, after which the curves begin to settle and flatten. Due to the intense build-up of heat up to this building height, significant microstructure gradients are predicted together with the specimen height. In Figure 3.13b, on the other hand, when measurements are taken before scanning a more stable temperature profile, heat can be observed up to 10 mm in height before abruptly accumulating. The excessive heat accumulation during remelting, as illustrated in Figure 3.13, has an impact on the cooling rates that the various build layers should undergo. Figure 3.14 depicts cooling rate curves derived from an indirect calculation based on mean surface temperature observations. The cooling rate gradient is most intense up

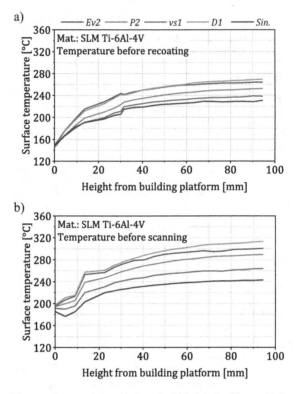

Figure 3.13 Mean surface temperature along build height in 90 mm high specimens: a) before recoating; b) before scanning

to a building height of 20 mm, following which the cooling rate stabilizes and flattens until the building height reaches 90 mm. Batch *Ev2* had the lowest residual porosity (based on μ-CT experiments according to section 3.3.1) and cooling rates along with the building height, indicating that cooling rate has an impact on melt pool stability and, as a result, build height. Residual stresses associated with SLM are a major drawback, limiting its application. Laser power, scanning stress-related distortions, and porosity all influence the accumulated residual stresses [182].

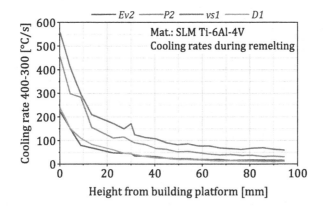

Figure 3.14 Cooling rate of 90 mm high specimens for selected parameter sets: *Ev2*, *P2*, *vs1*, *D1*

Functional grading

The findings reveal a link between remelting decrease of residual porosity and the impact of remelted microstructure on expected fatigue behavior. It is believed that by generating ultrafine microstructures near the surface where cracks originate from surface and subsurface porosity, they can create a graded structure from these remelting strategies. As a result, the likelihood of surface and subsurface fracture formation is reduced. Furthermore, remelting parameter settings that have been shown to improve ductility will be applied deeper in the specimen to generate a fracture entanglement process based on microstructure strengthening with enhanced ductility. Figure 3.15 depicts an illustrated schematic of such a suggested method based on Awd et al. [183], which includes two options. The cross-section of the specimen is not fully remelted on the left because we desire to keep the skin of a high proportion of α' phase in the subsurface layer of a specimen where no remelting or treatment is applied and the microstructure is expected to consist primarily of α' that is very acicular and fine and highly resistant to crack initiation in the as-built condition [184]. We compare this to the method in Figure 3.15 on the right, which uses the identical scanning sequence in the middle but remelts the skin to see how graded remelting affects surface and subsurface crack formation. However, dangers in the method are expected, which might lead to melting pool instabilities and defect formation at the interface or overlap of distinct parameter sets.

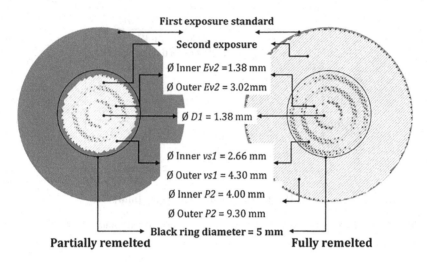

First exposure standard
Second exposure
Ø Inner *Ev2* =1.38 mm
Ø Outer *Ev2* = 3.02mm
Ø *D1* = 1.38 mm
Ø Inner *vs1* = 2.66 mm
Ø Outer *vs1* = 4.30 mm
Ø Inner *P2* = 4.00 mm
Ø Outer *P2* = 9.30 mm
Black ring diameter = 5 mm
Partially remelted   Fully remelted

Figure 3.15 Proposed schematic representation of the functional grading technique from the specimen's surface to its center [183]

3.3   Material Characterization on Multiscale

3.3.1   Structural Properties

Why microcomputed tomography (μ-CT)?
X-ray microcomputed tomography (μ-CT) is a non-destructive volume imaging technique that is specified in DIN EN ISO 15708 [185]. It allows the non-destructive volume inspection of materials and structures. μ-CT systems are composed of four primary components: an X-ray source, a detector, a specimen manipulator, and a reconstruction/visualization system, according to Figure 3.16. Three-dimensional μ-CT volumes enable the identification and quantification of volumetric features. The basis of μ-CT is based on the use of X-ray radiation, which is generated using an X-ray tube. At a transmission rate of 10—20%, the highest signal-to-noise ratio is reached. Extended exposure periods can compensate for reduced flux densities. Attenuation coefficients are determined by a variety of material-specific characteristics (density, atomic number, atomic weight) [186]. The produced volume is represented by voxels, which are displayed as gray values. Voxels can be thought of as 3D pixels. The object's picture sharpness is a result of the effective pixel size. The difference in the attenuation

coefficients, in conjunction with the resolution, affects the detail detectability of objects and features. μ-CT testing provides a 3D defect examination of materials and structures that is not achievable, or only partially possible, using planar light or scanning electron microscopy (SEM). This enables μ-CT inspection to be utilized for three-dimensional damage analysis. μ-CT systems and analytical techniques have enabled inspection times of less than 30 seconds to be achieved in some applications [187].

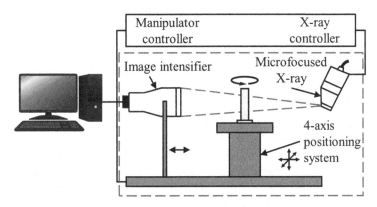

Figure 3.16 The experimental setup for X-ray scanning microcomputed tomography is depicted schematically [188]

Inhomogeneities in the structure of a μ-CT scan's sub-pixel or voxel area can have a substantial effect on the related mean linear attenuation coefficient μ. This phenomenon is referred known as the "partial volume effect," and it results in noticeable contrast in the blue scale depiction. Increase the signal-to-noise ratio by increasing the number of 2D μ-CT projections or viewing angles, such as in Figure 3.17. μ-CT-based defect and porosity analysis of additively manufactured lightweight materials have several benefits over more traditional procedures such as micrographs and section testing, ultrasonic testing, or density testing. Defect studies (e.g., porosity, inclusions, and damage) can be performed on the full sample volume or on specific regions of interest using appropriate algorithms. Increased uncertainty is related to porosity analysis using micrographs and section tests. This can result in undersized and, in the worst-case scenario, "unforeseen" component breakdown during service life [188]. Details with an edge length of two to three voxel size down to 3 μm can be identified with appropriate μ-CT system settings.

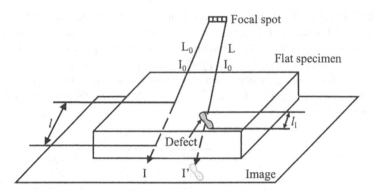

Figure 3.17 Schematic illustration of the influence of X-ray intensity on the identification of structural discontinuities [188]

By reconstructing a 2D image stack, a 3D representation of the scanned specimen can be created. The system gets a collection of projections or sets of angle views of the specimen to produce a reconstructed picture stack. Because the μ-CT method employs a linear detector, projections are stored as independent lines inside an image file called a sinogram. The vertical dimension of an image, i.e., the number of lines in sinogram images, equals the number of projections. The reconstruction of a sinogram produces a 2D slice across the object. X-rays should pass through the specimen to discover minute features. Beer's law can be used to calculate the intensity I of photons passed through the specimen [189]

$$I = I_0 e^{-\mu l} \tag{3.1}$$

$$I' \approx I_0 e^{[-\sum_k \mu(x_k)\Delta x]} \tag{3.2}$$

where x_k is a vertex along a trajectory $\mathbb{L}$ and $\mu(x_k)$ is the attenuation coefficient at x_k [188].

Experimental setup
X-ray microcomputed tomography (μ-CT) was used to study the effect of process variables on residual porosity. The statistical and 3D distribution of process-induced porosity in the gage length region was determined using a Nikon XT H 160 μ-CT. The μ-CT system is equipped with a 160 kV / 60 W microfocus

X-ray tube. A cannon produces a high-intensity X-ray beam that is focused into a specimen mounted on a revolving manipulator see Figure 3.18.

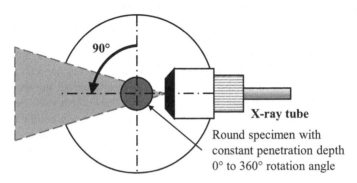

Figure 3.18 Illustration of the scanning process of the tested specimen geometries in the fatigue tests [188]

The change in the gray values of projected X-rays at the detector is utilized to generate a voxel-based volume in which structural discontinuities and defects are found and analyzed. The effect of scanning parameters on the generation of remnant porosity and the relationship between porosity and fatigue fractures following fatigue loading using the specified μ-CT facility were examined. The 1583 two-dimensional projections generated by X-ray scanning were used to construct a 3D data collection of volume elements (voxels). Pore identification was accomplished by viewing the reconstructed volumes using the application VGStudio Max 2.2 (Volume Graphics) and classifying them as defects if their grey value was smaller than a predefined threshold value. In this thesis, an effective pixel size of about 10 μm was used, along with the following μ-CT parameters: 115 kV, 59 A tube current, 2 fps exposure rate, 1583 projections, and 8-fold superimposition according to Table 3.3.

Table 3.3 Microcomputed tomography scanning parameters

Energy	Current	Power	Exposure	Pixel size	Projections
115 kV	59 μA	7.6 W	2 fps	10 μm	1583

Porosity analysis

Figure 3.19 illustrates a qualitative assessment of the defect status of AlSi10Mg. The scanning settings influence the stability of the melt pool and can be used to reduce Marangoni torques. As a consequence, the quantity of spherical metallurgical porosity can be decreased relative to previously published data [190]. This has a major influence on the shape of pores rather than their size, as demonstrated by the diameter Weibull distribution in Figure 3.20a, even though the total number of voids is much reduced. At 200 μm, the pore size reaches a maximum. AlSi10Mg has a relative density of 99.94%. It is clear from all the specimens that sphericity improves as the corresponding energy density increases. However, the overall number of pores is varied for the different specimens, but the relative frequency of sphericity is virtually identical for all scanned specimens, as seen in Figure 3.20c. A very tiny proportion of pores had sphericity less than 0.5.

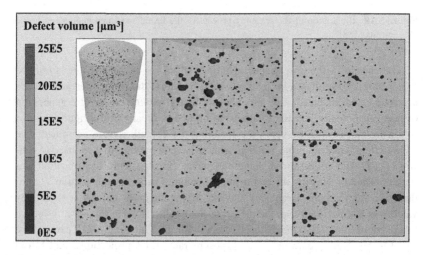

Figure 3.19 Three-dimensional computed tomographical maps of defect distribution of AlSi10Mg under SLM scanning

Due to the additional heat from the platform, the gas porosity increases somewhat, allowing a few or more to reach the computed tomography threshold of 10 μm. This is thought to be the explanation for the increased number of individual pores. The corresponding projected areas in the spatial directions are presented in Figure 3.20d to Figure 3.20f. When platform heating is missing, the number of pores is greater, particularly in the class 40 μm in diameter. Between 50 and

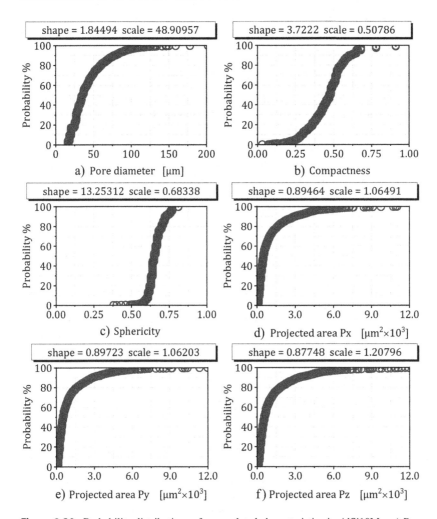

Figure 3.20   Probability distributions of pore-related characteristics in AlSi10Mg: a) Pore diameter; b) Compactness; c) Sphericity; Projected area in the d) x-direction; e) y-direction; f) z-direction

120 μm platform, heated specimens show a greater number of voids, according to Awd et al. [191]. Pore formation is a result of Marangoni convection and the torque that it generates. The intensity of the torque is related to the magnitude of the temperature differences in the melt pool area.

Due to centrifugal inertia, high-intensity torques move low-viscosity melt to the periphery, creating a hole in the melt pool's center [192]. The mechanism described here can be thought of as a metallurgical type of porosity that is fundamentally related to the thermodynamic circumstances that exist during consolidation when the applied local energy density exceeds the optimum. When the energy density provided is inadequate, partial fusion occurs in the bead, preventing the production of a solid track. The binding malfunction results in keyhole defects with sharp edges and high-stress concentration factors [193]. Keyhole porosity is more damaging to mechanical characteristics in general and specifically to fatigue strength, such as the notched defects seen in Figure 3.19. The number, position, and size of the pores can vary substantially depending on the direction of build-up. All specimens in this study are constructed in a 90° orientation. The scanning method is readily apparent from the alignment of the pores. Numerous voids of single regions are aligned equi-axially and clearly demonstrate the island scanning method, in which single quadratic sections in a single layer are melted sequentially. Considering similar investigation in the literature, there is a critical energy density point for this alloy, roughly 60 J/mm^3, at which the lowest pore fraction is obtained. Fracture surfaces exhibit a high concentration of unmelted powder, which results in local cracking [194].

Keyhole pores are formed as the scanning speed increases, accompanied by a decrease in metallurgical pores. A study demonstrated that the kind of pore generated during the SLM process is connected to the scanning time. Changing the scanning strategy to a double scan at high rates was an alternative to slowing down the scanning rate [193]. The observation that the biggest flaws in the SLM material were extremely anisotropic, being flat and disc-like in the xy-plane, is significant since the planes of these pores match the layer-wise manufacturing process planes. This might result in the development of innovative processing techniques that limit the formation of big pores [195]. Figure 3.21 depicts a qualitative assessment of Ti-6Al-4V's defect state. As indicated by the diameter distribution in Figure 3.22a, the stability of the melt pool has a significant effect on the form of pores rather than their size, even though the shape is usually flat and not spherical. The pore size reaches its greatest value at 180 μm. The shown Ti-6Al-4V specimen in Figure 3.21 has a 99.97% relative density. All specimens demonstrate that sphericity is less than that of AlSi10Mg. Although the total number of flaws varies among specimens, the relative frequency of sphericity is

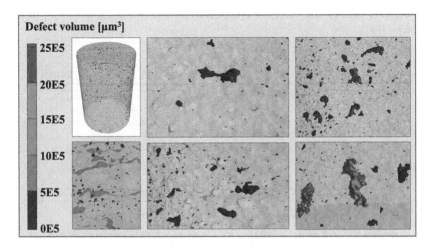

Figure 3.21   Three-dimensional computed tomographical maps of defect distribution of Ti-6Al-4V under SLM scanning

nearly equal across all scanned specimens, as seen in Figure 3.22c. In comparison to AlSi10Mg, a greater fraction of pores exhibited sphericity smaller than 0.5. The extra heat from the platform does not have the same impact as it does in AlSi10Mg because the applied 200 °C is insufficient for Ti-6Al-4V, which has a melting point of more than 2.5 times that of AlSi10Mg. As a result, platform heating has a lesser effect on Ti-6Al-4V than on AlSi10Mg. From Figure 3.22d to Figure 3.22f, the appropriate projected regions in the spatial directions are shown. According to Awd et al. [183], the proposed remelting procedures mentioned in section 3.2.2 were a successful strategy for decreasing porosity since nearly completely dense specimens could be created, as demonstrated by the μ-CT scans of the in situ heat-treated specimens. The corresponding projected areas in the spatial directions are presented in Figure 3.22d to Figure 3.22f. The application of μ-CT is strongly present in the literature. After HIP, the porosity shrank below the μ-CT limit (5 μm), and spherical gas pores containing argon gradually reappeared and grew in proportion to their original as-built size during high-temperature (annealing) treatments. No large irregular low-pressure pores reappeared [196]. Rapid expansion occurs due to vapor depression reflection of the processing laser and vapor depression oscillations [192]. Laser-metal interactions under such high temperatures are widely misunderstood. When the laser scan velocity changes, pores appear in the depressions and trap shielding gas. A mitigation strategy can eliminate pore formation and boosts the quality of melt tracks [197].

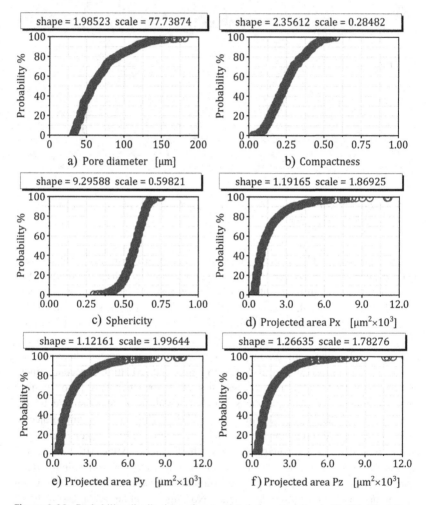

Figure 3.22 Probability distributions of pore-related characteristics in Ti-6Al-4V: a) Pore diameter; b) Compactness; c) Sphericity; Projected area in the d) x-direction; e) y-direction; f) z-direction

3.3.2 Microstructural Properties

AlSi10Mg and Ti-6Al-4V specimens were cut using the Labotom-3 cutting machine. The cut specimens measure 10 mm in height. Following the cutting procedure, the specimens must be hot embedded to facilitate grinding and polishing. To begin, there are two forms of embedding: hot and regular embedding. With normal embedding, we utilize epoxy resin and prepare the specimens for 48 hours; with hot embedding, preparation takes only 10 minutes. Additionally, because the embedding material is electrically conductive, they can be examined directly using SEM. The specimens are ground and polished to prepare the surface for further etching. The Tegramin-25 machine performs automatic grinding and polishing procedures by defining the duration, speed, and normal forces used. Grinding was carried out with grit paper ranging from 320 to 4000 grade, lubricated with water. Polishing was then performed from a height of 3 μm to 1 μm using diamond suspension. In general, etching metals refers to the process of degrading metal surfaces with a strong acid or alkaline. It is used to depict the phases seen in microstructures observed by light or scanning electron microscopy. The same etchant is utilized for AlSi10Mg and Ti-6Al-4V; however, the etching time is different. AlSi10Mg etching takes 10 s, but Ti-6Al-4V etching takes 30 s.

Light microscopy
AxioCam Zeiss light microscope is used for visualizing the microstructure of AlSi10Mg and Ti-6Al-4V for lower magnifications to 100X using polarized light. Figure 3.23 depicts the morphology of melt pools and melt pool boundaries of AlSi10Mg under SLM scanning using the light microscope in the z-direction. Melt pools' drop-like forms merge horizontally and vertically to produce a stable structure. There were no defects seen due to a lack of fusion. Randomly, small-scale porosity on the order of 40 μm in diameter was discovered. Pores observed in the non-platform heated 2D micrographs were similarly dispersed in their position. This is due to the additional thermal energy provided by platform heating, which increased densification and prevented the collapse of non-spherical gas pores. Porosity in Al-Si-Mg alloys can be classified into three types by Yang et al. [198]. No cavities developed randomly in any specimens were seen due to the presence of unmelted powder in cavities. This demonstrates that the scanning settings were enough to generate a solid fusion without heating the platform. On the other hand, small-scale gas voids developed at the boundaries because of melt overlaps produced by the excessively high energy density.

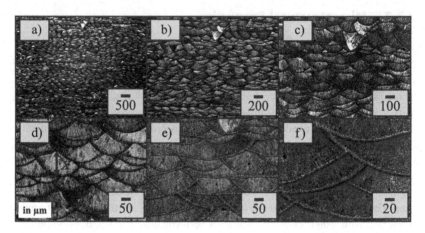

Figure 3.23 Morphology of melt pools and melt pool boundaries of AlSi10Mg under selective laser melting scanning using the light microscope in the z-direction

The blue X in Figure 3.23d represents the melt pool height. AlSi10Mg alloy contains nanocrystalline dendritic silicon particles that contribute to its exceptional hardness, strength, and impact toughness. While flaws parallel to the loading axis (or parallel to the fracture plane) are particularly detrimental, when oriented in a favorable direction, the alloy obtained literature-similar strength and ductility that correspond to alloys generated using SLM [199]. Using microstructure analysis, components made from stock powders and those made with nanoparticle-functionalized powders differ markedly. Stocks of 7075 and 6061 have massive columnar grains parallel to the build direction, with fractures extending across many build layers. As with earlier attempts to AM wrought aluminum alloys, the additive technique generates a strong, directed heat flux [200]. Researchers studied the creation of single scan tracks and layers, which are fundamental building blocks for larger consolidated SLM components. With increasing laser scan speed, the diameters of lines and tracks produced by SLM decreased linearly. There was no indication of porosity in track or layer cross-sections, suggesting that pores emerge during layer accumulation during multilayer processing [201].

Due to the extremely fine microstructure and homogeneous distribution of the alloying constituents, the as-built material exhibited consistent nano-hardness throughout. Although heat treatment reduced the material's compressive yield strength, it remained more than the die-cast counterpart. Both the as-built and heat-treated materials proved their capacity to sustain high compressive stresses by barreling and buckling during compression without fracturing [202]. As-fabricated samples exhibit a melt pool pattern because of the SLM scan strategy, with the melt pools consisting of a dendritic cell network with 1 μm-sized cells. Columnar grains were identified running parallel to the construction direction, with equiaxed grains discovered in the cross-section. The breadth and diameter of the grains are around 5 μm. This was related to the rapid cooling rates encountered during SLM. Additionally, undercooling and recalescence are highly likely to be significant with HIPing, which was discovered to fill irregular-shaped voids, although oxide layers remained in the microstructure [203]. For the powder to be dried, less laser power is employed. An internal laser drying process can achieve up to a 90% decrease in moisture and pore density [204].

Figure 3.24 Morphology of prior β grains and $\alpha + \beta$ regions of Ti-6Al-4V under selective laser melting scanning using the light microscope in the z-direction

The bulk microstructure of Ti-6Al-4V specimens is characterized by columnar prior β grains aligned in the build direction and continuous colonies of Widmanstaetten structure ($\alpha + \beta$) along with the retained phase, see Figure 3.24. Columnar prior grains are a direct result of directional solidification and significant temperature gradients in the direction of the building. In titanium alloys, microstructure development is regulated by a complicated interaction between the cooling rate from a high-temperature single phase to multiple phases and the crystallographic orientation of the parent phase. Figure 3.24f shows an enlarged view of a prior grain, demonstrating the change in phase morphology at and around the prior grain border. A favored orientation connection results in the development of 40—60 μm wide lamellar colonies. The decrease in the average lath width at the junction can be attributable to the increased cooling rate during each subsequent layer deposition. The width of the prior β is statistically measured as shown by the variable y in Figure 3.24a.

Microstructure and phase composition in SLM Ti-6Al-4V significantly influences its mechanical properties. Lattice defects play a role in the precipitation of the microstructure [205]. A study showed that commercial purity of titanium implants with varying porosities contained fine grains and intermetallic Fe-Ti compounds with a 40 nm diameter. They also have comparable yield strength and low elastic bone modulus [206]. The microstructure evolution that occurs following SLM is strongly influenced by the initial acicular martensite stability. The heating temperature had a much greater effect on alloy strength and fracture strain than the cooling rate. A post-SLM heat treatment at 850 °C or higher can result in an increase in fracture strain to that of the forged counterpart but at the expense of yield strength and ultimate compressive strength [207]. Within the massive phase grains, ultrafine $\alpha + \beta$ lamellar structures were formed by $\alpha' \rightarrow \alpha + \beta$ in situ decomposition. These structures can be used to enhance the ductility and yield strength of EBM-fabricated Ti-6Al-4V. The effect can be mimicked in SLM by in situ remelting [183]. Findings can stimulate the design/manufacture of new microstructures for enhanced properties in AM-based Ti alloys [208]. Optical and electron microscope tests showed a microstructure that is peculiar to martensite. The progression of maximal tensile and compressive stresses showed an unusual phenomenon: asymmetric cyclic softening. This behavior was detected exclusively in SLM Ti-6Al-4V, although its wrought counterpart corroborated prior observations [209].

Scanning electron microscopy (SEM)
Scanning electron microscopy (SEM) analysis was performed using a Mira 3 XMU, Tescan, Brno, Czech Republic, and energy-dispersive X-ray spectroscopy (EDS)-detector (Octane Pro, EDAX/Ametek, Berwyn, PA, USA). SEM was used to magnify the melt tracks. This scale shows smaller scale porosity better. They are strewn over the melt puddles. They were rarely encountered along the edges. Thus, laser scan overlaps increased energy density but not excessively. It finds that the development of small-scale porosity in melt pools is distinct from Aboulkhair et al. [193]'s balling and Marangoni force process. Figure 3.25 illustrates high magnification scales along the build direction. In the

tensile and uniaxial fatigue tests, the shown plane is the primary loading direction. Figure 3.25b and Figure 3.25c illustrate the melt pool center and tracks in the z-direction. Compared to xy-direction investigations in Awd et al. [191] there is less inhomogeneity in the z-direction. The melt pool center has small-scale porosity of < 10 μm, as shown in Figure 3.25a.

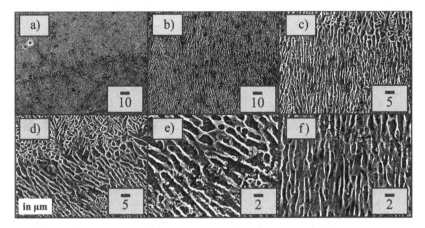

Figure 3.25 Multiscale morphological characterization of the eutectic phase of AlSi10Mg under selective laser melting scanning at the melt pool center and boundary using high magnification scanning electron microscope

The presented unique microstructure of SLM AlSi10Mg has a positive effect on strain rate dependency. For example, dynamic loading of a direct metal laser sintering (DMLS)-AlSi10Mg 200 °C at two strain rates led to dynamic recovery due to an ultrafine substructure developed inside the Al dendrites at a rate of 1680 s^{-1} in the presence of Al dynamic recovery (DRV). At 4300 s^{-1}, continuous dynamic recrystallization (DRX) occurred, which was followed by DRV inside the DRX grains, which resulted in the formation of nanoscale subgrains in the DRX grains [210]. A team from Monash University found a new technique to refine columnar grains separated by sub-micron equiaxed grains near the melt pool borders. Adding 1.08 wt.% Sc to an alloy generated by SLM alters the very coarse columnar grain structure. The latter nucleates predominantly from the remelting zone Al3Sc particles. A nearly equiaxed grain structure was produced by increasing the volumetric energy density applied from 77.10 J/mm^3 to 154.20 J/mm^3 and platform temperature from 35 °C to 200 °C [211].

The fatigue characteristics of allotropic Ti alloys are more susceptible to microstructural and textural influences than the fatigue qualities of alloys possessing isotopic FCC or BCC crystal structures, respectively [212]. In Figure 3.26, the SEM micrography has an acicular α' martensite microstructure, which accounts for most of its composition. In case of no melting overlaps, the prior β grain boundaries are clearly visible. This is because the previous layers serve as a thermal sink, resulting in greater cooling rates in the initial layers and, as a result, lower lamellae thickness.

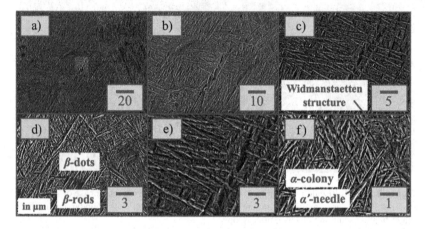

Figure 3.26 The distribution of α' and $\alpha+\beta$ phases within a prior β at the scanning electron microscope under selective laser melting scanning for Ti-6Al-4V

As seen in Figure 3.26d and Figure 3.26f, a basketweave structure can be found. This means that to avoid brittleness, the laser power for the overlapping scan should be lowered suitably as well. The melt tracks were characterized by the presence of several parallel heat-affected zones at some locations. As illustrated in Figure 3.26a, these spots represent regions of microstructure transition, which results in a distinctive hue when the specimen is seen in light microscopy. In general, the SLM process involves the heating and cooling of the material on a cyclical basis. Heat treatment in situ occurs when additional laser scans are performed on the solidified material after it has been heated to a certain temperature by the lasers. Due to microstructural heterogeneity, oxide films that form due to thermal degradation and moisture still remaining in the powder can increasing pitting corrosion resistance during fatigue loading [213].

In EBM-printed Ti-6Al-4V block samples with different printing thicknesses, the evolution of the martensitic transition and/or interface of a general phase-change sequence was observed. Higher in-fill hatching thickness leads to greater partitioning ratios and fraction volume. It can result in an increased lattice mismatch and weaker α/β interfaces, which can explain why strength declines as thickness increases [214]. In both as-built and stress-relieved samples, X-ray diffraction (XRD) and transmission electron microscopy (TEM) examination showed a totally martensitic structure with a high dislocation density and stacking faults. Twining is said to arise because of the high temperatures experienced during manufacturing [215]. SLM enables in situ α' martensite breakdown during the building of Ti-6Al-4V, converting the undesirable martensitic structure to an extremely lamellar ($\alpha + \beta$) structure. The heat treatment time accrued because of the thermal cycling effect during SLM has been shown to be adequate to convert acicular martensite to lamellar [216]. Temperature, time, and cooling rate all influenced phase precipitates with an original microstructure at the α' borders of Ti-6Al-4V. At maximum temperatures below the β transus, the combination of pre- and post-precursor grains prevents grain development, and columnar prior grains remain evident after cooling. With increasing temperature, yield strength σ_y and ultimate tensile strength (UTS) decreases while the fracture strain increases due to the tiny needles transforming into a coarser mixture of α and β. The best overall results are achieved after two hours at 850 °C followed by furnace cooling or after one hour at 940 °C followed by air cooling and tempering [217].

3.3.3 Quasi-static Properties

Tensile testing
Quasi-static tensile testing was performed on the machined specimens using an Instron 3369 system in compliance with ISO 6892–1:2009 standards [218]. To monitor the development of strain during the test and to compute the findings, an extensometer with a gauge length of 10 mm was utilized. Among other characteristics, it has a load cell with a nominal force capacity of 50 kN. An Instron type 2620–603 extensometer with an initial gauge length of l_o=10 mm, and a measurement range of $\Delta l = \pm 1$ mm was used to monitor the test (class 0.2 according to DIN EN ISO 9513 [219]). The specimen geometry is shown in Figure 3.31. The stress-strain graphs and computation of the characteristic values were generated in a spreadsheet tool called OriginLab utilizing the raw data.

The stress-strain curves for AlSi10Mg are presented in Figure 3.27. As expected, SLM AlSi10Mg exhibited greater yield strength, tensile strength, and fracture strain than the cast counterpart [220]. The curves depicted are three tests of the same batch, with the representative curve highlighted in dark blue. The yield strength did not vary substantially statistically. Otherwise, the average of three tests indicates that AlSi10Mg has a tensile strength of 451.07 MPa, which is higher than the non-platform heated variant as reported by Awd et al. [191]. On the contrary, the statistical fracture strain for platform-heated specimens was higher. Due to the more homogeneous structure created because of platform heating, the characteristics are more consistent and dependable with lower variance. AlSi10Mg has a tensile strength of 396 ± 8 MPa with a fracture strain of 3.47 ± 0.60 [10^{-2}], according to Kempen et al. [221]. Tensile strength was increased because of the extremely small cellular structure and supersaturation of Si particles. Tang et al. [222] reported a compressive strength of 287 MPa for recycled AlSi10Mg powder as a result of oxide and impurity development.

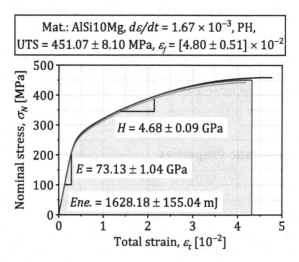

Figure 3.27 Quasi-static tensile properties of AlSi10Mg based on three tests (Energy · 1E2)

Characteristics of the AlSi10Mg alloy processed by SLM in hot and cold platform processing were considered, among other strategies, to assess the aging response of the alloy's mechanical properties. In particular, a hot platform was used in order to achieve improved mechanical properties on a higher level of hardening in situ aging response [223]. A team of scientists at the University of Tennessee investigated Al-6061 structures in x- and z-axes. EDS detected no oxygen peaks, and there was no obvious interface line. As a result of strain localization, micro-voids develop at the contact and combine to nucleate fractures that propagate across the interface, resulting in mode-I fracture without ductility [224].

The stress-strain curves for Ti-6Al-4V are presented in Figure 3.28, demonstrating that SLM Ti-6Al-4V with the employed parameter set has a greater tensile strength of 1280.59 MPa and a higher yield strength than wire + arc additively manufactured (WAAM) Ti-6Al-4V [56]. Platform heating had no effect on the elastic behavior, according to Awd et al. [183]. However, as indicated in the statistical results of Figure 3.28, hardening in the plastic regime of Ti-6Al-4V specimens was inversely related to AlSi10Mg in Figure 3.27 as indicated by the hardening modulus H, resulting in somewhat first evidence of softening behavior in cyclic loading. It is worth mentioning that the energy (*Ene.*) absorbed by the Ti-6Al-4V specimens is more than 8 times more than AlSi10Mg specimens. Both alloys achieved a higher strength than their conventional counterparts [225].

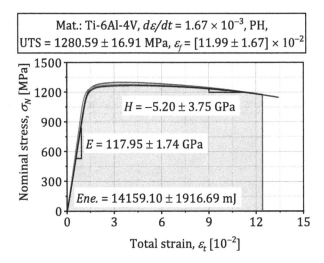

Figure 3.28 Quasi-static tensile properties of Ti-6Al-4V based on three tests (Energy · 1E2)

Non-spherical pores were identified in laser-processed Ti-6Al-4V specimens and were primarily attributed to inadequate remelting of the preceding layer's particular localized surface areas and insufficient feeding of molten metal to solidification fronts. HIPing reportedly successfully removed virtually all porosity, transforming the martensite structure into $\alpha + \beta$ phases. This resulted in a substantial improvement in ductility but reduced strength [226]. The direction in which SLM Ti-6Al-4V is oriented influences its ductility. This is due to the orientation of the prior-grain boundaries with respect to the external axial loading direction. The microstructure's directionality influences the fracture processes and crack propagation in the components [227].

Microhardness testing
Hardness is defined as a material's resistance to indentation. The permanent indentation depth is, therefore, the measure of this resistance. A large depth length results in a soft substance. The material will resist indentation and have a high hardness if the depth is shorter. The Vickers hardness test determines the Vickers pyramid indenter force for AlSi10Mg (HV0.2 $\rightarrow$ kgf/mm^2) and Ti-6Al-4V (HV1.0 $\rightarrow$ kgf/mm^2). Pressure is determined by force applied to the indentation's surface area, not by the weight delivered. For each microscopic section, an average of 30 measurements is taken to establish how hardness distributes itself across a specimen surface area and the average hardness value for each specimen. The automatic micro-Vickers hardness tester Shimadzu HMV-G was used for creating hardness distributions of AlSi10Mg and Ti-6Al-4V.

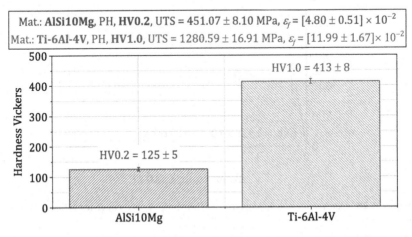

Figure 3.29 Comparison of microhardness using Vickers indenter between AlSi10Mg and Ti-6Al-4V

The average hardness measured in the z-direction for AlSi10Mg and Ti-6Al-4V is 125 and 413, respectively; see Figure 3.29. Microstructure results of SLM-processed Al-Si alloys with extra-high silicon content showed that the primary silicon phase is much smaller than that of conventional casted samples. Therefore, hardness can reach up to 188 HV [228]. Research has shown that adding nickel significantly enhanced the toughness of the Al-Si-Ni alloy. This was also proven using nano-indentation measures and microstructure examinations. The researchers discovered that, even when Al3Ni agglomerates are present, SLM provides for excellent nickel dispersion inside the alloy [229]. The inclusion of Si can lower the melting point of the alloy, increasing its fluidity. With the addition of 4 wt.% Si, a new eutectic is formed, and considerable grain refinement occurs, preventing the development and spread of fractures. The unique microstructure of the Al7075 + 4 wt.% Si treated by SLM, coupled with the appropriate age treatment, results in hardness values (171 ± 4) HV equivalent to conventional Al7075 with T6 treatment 175 HV [230].

A nearly flawless Ti-6Al-4V had a microhardness level of 450 HV [231]. A 1200 °C treatment dissolved the structure entirely and replaced it with an equiaxed structure. This significantly enhanced the average hardness of the specimens [232]. Tensile, fatigue, and hardness properties were substantially reduced when the defects reached 5% in EBM [233]. However, hardness in as-built conditions was found to be higher than that of the wrought bar in both longitudinal and transverse orientations. [234]. Microhardness was reported to increase as scanning speed increased [235]. With increasing Mo content, grain size decreased, while equiaxed grain layer thickness increased. This indicates a tendency toward columnar-to-equiaxed transition occurs as Mo increases, which influences hardness positively [236]. It was notably shown that the hardness could range from HRC 37 to HRC 42 within a dimensional range of 4 μm without any systematic changes to the construction parameters and to attain a maximum HRC 50 [225]. TiB/Ti-6Al-4V composites had superior nano hardness compared to the sintering counterpart. The in situ synthesized TiB reinforcement noticeably improves the wear resistance properties, resulting in an approximately 2-fold increase in the wear resistance [237].

Instrumented indentation
Elastic modulus is a fundamental mechanical parameter for determining the microstructural strength of a material. Elastic modulus is determined using the depth-sensitive indentation technique. Not only does the surface deform as the notch depth increases in the microhardness test, but the subsurface commonly

exhibits elastic and plastic deformations as well [238]. To determine a microstructure elastic modulus, a general rule of thumb is to keep the indentation depth to less than 10% of the specimen thickness [239]. A technique for determining the elastic modulus of microstructures was applied, i.e., a thorough knowledge of the subsurface effect on mechanical performance findings such as indentation depth and contact area [240].

Using the ultra-micro instrumented indentation test, the elastic modulus is determined by the following procedure [241]

$$E^{SI} = \frac{\sqrt{\pi}}{2} \frac{S}{\sqrt{A}} \tag{3.3}$$

where A is the notch contact area under the maximum load, and S is the slope of the unloading curve. E^{SI} is a 'specimen + indenter' modulus. The elastic modulus E of the sample depends on the elastic modulus of the surface and subsurface (E_f and E_s, respectively), and is determined using the following definition

$$\frac{1}{E^{SI}} = \frac{1}{E^*} + \frac{1}{E_i^*} \tag{3.4}$$

With

$$E^* = E/\left(1 - \upsilon^2\right) \, and \, E_i^* = E_i/\left(1 - \upsilon_i^2\right) \tag{3.5}$$

where E^* and E_i^* are the elastic moduli of reduction, and υ and υ_i are the Poisson ratios of the specimen and the indenter, respectively. In the case of an elastic homogeneous material, the surface and the subsurface have the same reduced elastic modulus, that is $E_f^* = E_s^*$ and E^* represent the value of the sample reduced elastic modulus: $E^* = E_f^* = E_s^*$.

True projected contact area
The dependency of hardness and elastic properties on the true contact area is discussed as follows [242–244]

$$H = \frac{P}{A_c} \tag{3.6}$$

$$\frac{1 - \upsilon^2}{E} = \frac{1}{E_c^*} - \frac{1 - \upsilon_i^2}{E_i} \tag{3.7}$$

$$E_c^* = \frac{S}{2}\sqrt{\frac{\pi}{A_c}} \tag{3.8}$$

H is the hardness modulus and E is the elastic modulus of the material. In Equation 3.6 to Equation 3.8, P is the applied load, E_c^* is the reduced contact modulus between the tip and the specimen, A_c is the projected contact area. In Equation 3.7 and Equation 3.8 except for A_c all parameters are known or measured by the indentation test. For a perfect pyramid penetrator, the projected contact area is related to the contact depth h_c through the geometry of the tip

$$A_c = \pi \cdot tan^2\theta \cdot h_c^2 \tag{3.9}$$

where θ is the equivalent half-angle at the vertex of the tip. The estimated contact area will be reduced by the blunting effect of the tip at depths less than a few hundred nanometers. Oliver and Pharr devised a complicated tip area function that accounts for tip blunting by using several variables [245]. Due to the complicated procedures required to obtain the coefficients, less laborious models have been developed. For example, Loubet et al. recommended to explain the tip defect by adding a small height to the offset signal [246]. Other researchers advise changing the square root of the predicted contact using coefficients as a function of the linear h_c. These qualities are easiest to achieve experimentally at depths larger than 200 nm [247, 248]. For shallower depths, other models using power laws or exponential functions have been used [249, 250]. Of particular interest is Chicot et al. [251] an area function using linear and exponential terms is proposed to explain tip defects, which can be accurately determined using SEM images. The link between the contact depth and the indentation depth (as measured by indentation technology) is not straightforward, since the eventual accumulation or sinking that occurs throughout the indentation process cannot be identified only by monitoring tip displacement. Numerous hypotheses have been proposed to account for this occurrence [245]

$$h_c = h - \varepsilon\frac{P}{S} \tag{3.10}$$

where $\varepsilon = 0.75$ is a geometrical factor for a pyramidal indenter. P is the maximum load. S is the slope of the unloading curve. h is the reading depth on the experiment values. This method does not consider the pile-up simply because in Equation 3.10, h_c cannot be greater than h. The measurements were carried out on a Shimadzu dynamic ultra-microhardness tester DUH-211 for AlSi10Mg

a)

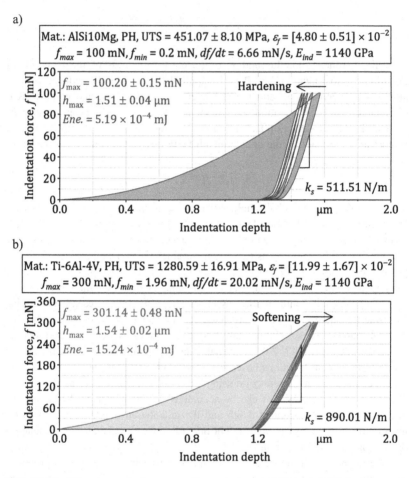

b)

Figure 3.30 Ultramicroscale flow properties under quasi-static repeated indentations: a) AlSi10Mg; b) Ti-6Al-4V

and Ti-6Al-4V. Here, an example of elastic modulus determination is shown for AlSi10Mg (Figure 3.30a) such that a Vickers indenter type has been used with an elastic modulus of $E_{ind} = 1.14 \cdot 10^3$ GPa and a Poisson's ratio of 0.07. A semi-apex angle as in Equation 3.9 is fixed to $68°$. To calculate h_c, the maximum depth of the first unloading is considered, which is 1.3418 µm. Since P corresponds to the maximum load of the first unloading also, the elastic modulus in the case of Figure 3.30a is ~ 76.70 GPa for AlSi10Mg. For Ti-6Al-4V, the elastic modulus is ~ 124.80 GPa based on Figure 3.30b. Both values were higher than the obtained elastic modulus from the standard tensile test. Additionally, the mechanical properties of functionally graded binary alloys were characterized by the same method. The advantage was that influence of the process parameters on the mechanical properties could be measured quickly and efficiently [252]. An instrumented indentation technique was used in deep neural networks, including a multi-fidelity approach to extract the elastoplastic characteristics of metals and alloys from the data collected [253]. Additionally, a constraint factor was used to translate hardness to yield strength [254]. Instrumented indentation is used in this study to determine the influence of process parameters on the flow properties of the microstructure, which is isolated from the effects of the defects.

3.3.4 Fatigue Strength

Servohydraulic testing
Fatigue testing on AlSi10Mg was conducted using an Instron 8872 servohydraulic test system fitted with a ± 10 kN load cell. For Ti-6Al-4V specimens, a Schenck-type PC63M system with a nominal force of ± 63 kN was employed. The measurement and control system on the testing systems is an Instron type 8800. All tests were conducted with a stress ratio $R = -1$, therefore a totally reversed loading. The specimens were clamped in hydraulic-controlled clamps to provide uniform clamping pressure and to eliminate any clamping force influence. Wave Matrix TM, the associated software, is used to operate the system during testing. The axial stress is delivered in the servohydraulic testing system via an electro-hydraulic control valve (referred to as a servo valve), which regulates the inflow and outflow of the two-chamber hydraulic cylinder. An electro-hydraulic control valve is a critical component of the system because it regulates the intake of hydraulic oil from a pressure reservoir into the test cylinder and the return flow of the oil to the hydraulic power unit's reservoir. The fatigue tests were conducted, and the measurement data were captured using Wave Matrix V software. Extensometer type 2620–603 (class 0.2 according to DIN 9513 [219]) with a gauge length of 10 mm was used to perform testing at a frequency of 20 Hz for AlSi10Mg and 5 Hz for Ti-6Al-4V using test specimens with the geometry indicated in Figure 3.31. In the testing optimization phase, the load increase tests (LIT) were employed, which significantly decreased the time to select the stress level at which a constant stress amplitude test (CAT) should be carried out.

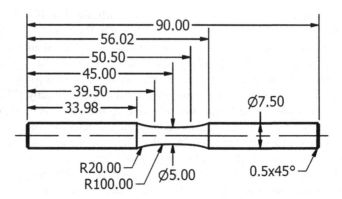

Figure 3.31 Technical drawing of specimen geometry used during servohydraulic fatigue tests for AlSi10Mg and Ti-6Al-4V

The stress level associated with the non-monotonic growth of plastic strain amplitude might be regarded as the crucial stress range for which additional testing must be conducted. A stress level within this range was chosen, and several constant amplitude tests were conducted to establish the mean and scatter of fatigue life for the two alloys, allowing for the identification of distinguishable failure mechanisms. In Figure 3.32, cyclic deformation under incremental loading is shown for AlSi10Mg. The control parameter is the stress amplitude. The response parameters are the total mean strain, the mean plastic strain, the plastic strain amplitude, and the dynamic elastic modulus. The test starts at a stress amplitude of 30 MPa, and the stress increases at a rate of 10 MPa/10^4 cycles. The total mean strain and the plastic mean strain increase at a higher rate at the beginning of the test. Then the rate decreases after 100 MPa. A sharp increase then follows at around 130 MPa. The plastic strain amplitude increases slowly at the beginning of the test until 3E4 cycles. Then the rate at which plastic strain accumulates increases gradually until 165 MPa. After reaching this load level, the specimen starts to harden towards fracture, as evident by the slight gradual decrease of plastic strain amplitude. The dynamic elastic modulus of the specimen shows a stable behavior until 1.5E4 cycles, after which the specimen starts to stiffen until reaching its half-life. Gradual then abrupt decline of specimen stiffness commences towards complete failure. Worth mentioning that the highest stress amplitude reached here is around 30 MPa higher than that of AlSi12 subjected to the same test in [255].

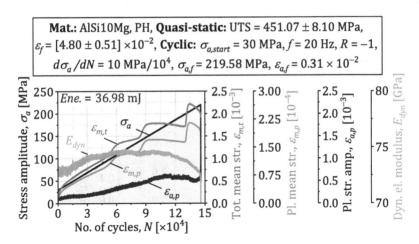

Figure 3.32 Cyclic damage accumulation in a load increase test evident by damage state parameters in AlSi10Mg (Energy · 1E2)

Figure 3.33 depicts the same damage parameters for Ti-6Al-4V, whereas seen stress amplitude starts at 150 MPa of controlled stress. The stress amplitude is increased at a rate of 20 MPa/10^3 cycles. Contrary to AlSi10Mg, the total mean strain and mean plastic strain increased and decreased respectively at the beginning of the test. Total mean strain then will reach a horizontal plateau until a quarter of the life of the specimen, which will begin then to decrease, indicating the activation of cyclic creeping sensitivity at stress amplitudes higher than 400 MPa. The plastic strain amplitude increases monotonically at the beginning of the test until around 450 MPa of stress amplitude which means a direct uniform load and deformation sensitivity relationship. Higher than 450 MPa of stress amplitude, plastic strain amplitude drops slightly and gains a parabolic relationship with loading, which indicates the engagement of further interactions of cyclic deformation mechanisms such as severe strain localizations, microstructural hardening, and formation of a multitude of cracks.

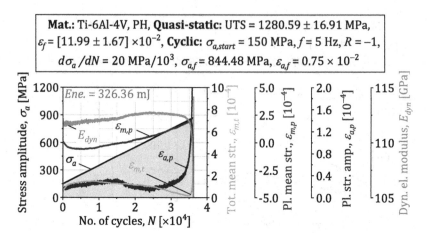

Figure 3.33 Cyclic damage accumulation in a load increase test evident by damage state parameters in Ti-6Al-4V (Energy · 1E2)

The significance of the load increase tests is their capability of capturing the multiscale fatigue damage regimes as described in the literature [256]. Microstructural transformations in the bulk of the material occur early on because of cyclic deformation, eventually leading to some type of strain localization at microstructural discontinuities. This is presented in the LIT by a slow rate of plastic strain accumulation at the beginning of the tests. In a subsequent stage, a crack nucleates which leads to significant compliance of the fatigue specimen that is reflected microscopically by a higher rate of accumulation of plastic strain. The other parameters which are measured, such as total mean strain and plastic mean strain, indicate the amount of cyclic creeping induced on the specimen.

Therefore, comparing the LIT stress-strain result to the tensile stress-strain result indicates the superiority of one alloying system over the other to develop hardening as loading cycles are repeatedly applied. In Figure 3.34, the alloying system of AlSi10Mg develops cyclic hardening in comparison to the $\alpha + \beta$ nature of Ti-6Al-4V. Based on the cyclic deformation behavior of LIT of AlSi10Mg in Figure 3.32, the range of constant amplitude tests (CAT) for AlSi10Mg was determined to be between 100 and 140 MPa. Figure 3.35 shows three CATs at 100, 120, and 140 MPa, respectively. At 100 and 120 MPa, the specimens show a continuous decline of elastic and plastic strains, including the amplitude and rise for the mean components indicating hardening behavior in the isotropic and cyclic creeping in the kinematic characteristics of the deformation for a completely

a)

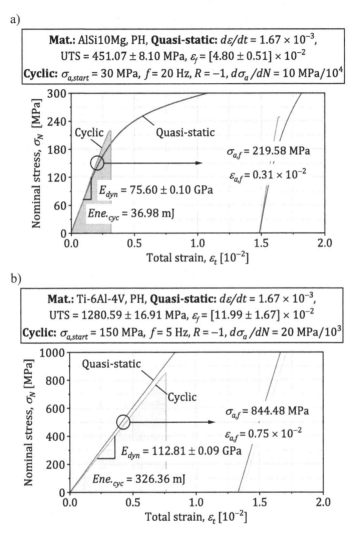

Mat.: AlSi10Mg, PH, **Quasi-static:** $d\varepsilon/dt = 1.67 \times 10^{-3}$,
UTS = 451.07 ± 8.10 MPa, $\varepsilon_f = [4.80 \pm 0.51] \times 10^{-2}$
Cyclic: $\sigma_{a,start} = 30$ MPa, $f = 20$ Hz, $R = -1$, $d\sigma_a/dN = 10$ MPa/10^4

Cyclic   Quasi-static

$\sigma_{a,f} = 219.58$ MPa

$\varepsilon_{a,f} = 0.31 \times 10^{-2}$

$E_{dyn} = 75.60 \pm 0.10$ GPa

$Ene._{cyc} = 36.98$ mJ

b)

Mat.: Ti-6Al-4V, PH, **Quasi-static:** $d\varepsilon/dt = 1.67 \times 10^{-3}$,
UTS = 1280.59 ± 16.91 MPa, $\varepsilon_f = [11.99 \pm 1.67] \times 10^{-2}$
Cyclic: $\sigma_{a,start} = 150$ MPa, $f = 5$ Hz, $R = -1$, $d\sigma_a/dN = 20$ MPa/10^3

Quasi-static

Cyclic

$\sigma_{a,f} = 844.48$ MPa

$\varepsilon_{a,f} = 0.75 \times 10^{-2}$

$E_{dyn} = 112.81 \pm 0.09$ GPa

$Ene._{cyc} = 326.36$ mJ

Figure 3.34 Comparison of quasi-static and cyclic flow behavior: a) AlSi10Mg; b) Ti-6Al-4V (Energy · 1E2)

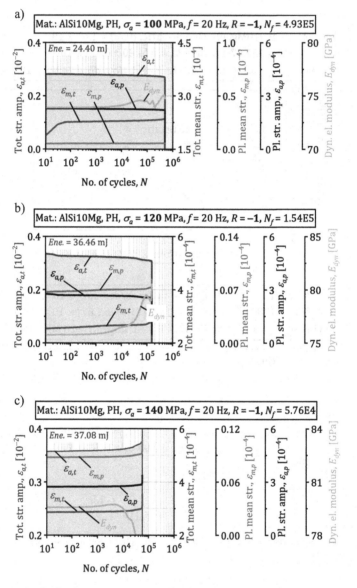

Figure 3.35 Cyclic damage parameters of AlSi10Mg in constant amplitude testing: a) 100 MPa; b) 120 MPa; c) 140 MPa of stress amplitude (Energy · 1E2)

reversed loading. The dynamic elastic modulus at the two load levels begins stable at the start of the test and then increases in an unstable manner at the end. The instability is a result of the build-up of internal cracks. On the other hand, during the 140 MPa test, the strain components are building up proportionally, indicating softening with degradation of the dynamic elastic modulus. The results suggest a change in the fatigue damage mechanism at higher loads. Based on the LIT of Ti-6Al-4V in Figure 3.33, the stress amplitudes of CATs were selected as 600 and 800 MPa. In Figure 3.36, the cyclic deformation curves are shown with the elastic and plastic strains, including the amplitude and mean components. Although the applied stresses are constant, the response of Ti-6Al-4V is highly nonlinear in comparison to AlSi10Mg in Figure 3.35. This is ought to the higher complexity of the $\alpha + \beta$ microstructure of Ti-6Al-4V which consists of prior β grains at the mesoscale and within resides Widmanstaetten structures with α' martensite and further α colonies with a balance that depends on the thermal history of the melt track. At both stress amplitudes softening is observed, which is confirmed for the HCF regime in the literature and explained by the fact that the higher the proportion of α is, the higher the slip distance a colony can accommodate. Thus, more capacity for cyclic straining and plastic strain accumulation [212]. However, from Figure 3.36, the amount of total strain is significantly decreased as the test progresses; thus, the total influence is a slight stiffening of the specimens overall, as evident by the dynamic elastic modulus. Since titanium alloys exhibit a rate-dependent behavior, the amount of strain accumulation ought to be different when testing with different frequencies [256]. Additionally, a significant amount of ratcheting and kinematic softening behavior was observed at 800 MPa, as shown from the mean components of strain. The amount of kinematic softening at 600 MPa is significantly less.

The deformation behavior indicates two important conclusions, which are the strong influence of the microstructure on the nonlinearity of the behavior, which will suggest a strain rate sensitivity. In Figure 3.37, the S-N relation in SLM AlSi10Mg of this analysis is shown where it is characterized by high scatter even in the early HCF regime. The highest life at 140 MPa is almost 10 times higher than the lowest life. The amount of scattering does not show a trend with decreasing stress amplitude. The fatigue strength here is slightly lower than that is reported in the literature on cast A356-T6 [257]. The non-homogeneity of the structure on the meso and microscale with a wide variability of features is likely highly responsible, as will be discussed in the next lines. It was reported that fatigue failures arose from the surface and sub-surface flaws (maximum surface distance of 275 μm) in all studied instances. Clusters of pores caused fatigue

a)

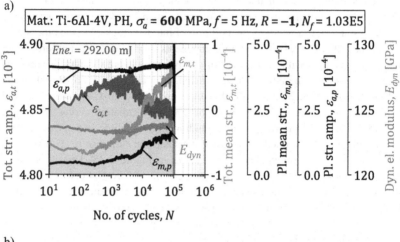

b)

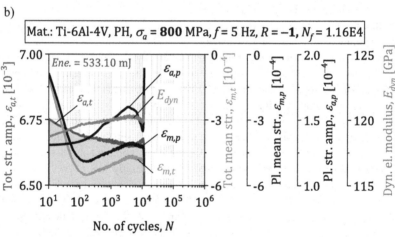

Figure 3.36 Cyclic damage parameters of Ti-6Al-4V in constant amplitude testing: a) 600 MPa; b) 800 MPa of stress amplitude (Energy · 1E2)

breakdowns in most situations (67%) [258]. The area near the surface was determined to be critical for VHCF response, whereas bigger non-critical interior flaws were identified further from the outside surface.

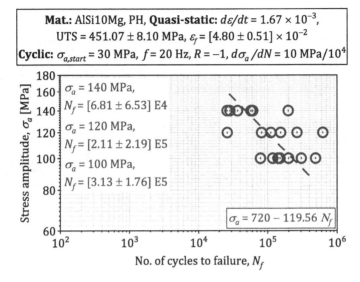

Mat.: AlSi10Mg, PH, **Quasi-static:** $d\varepsilon/dt = 1.67 \times 10^{-3}$,
UTS $= 451.07 \pm 8.10$ MPa, $\varepsilon_f = [4.80 \pm 0.51] \times 10^{-2}$
Cyclic: $\sigma_{a,start} = 30$ MPa, $f = 20$ Hz, $R = -1$, $d\sigma_a/dN = 10$ MPa/10^4

$\sigma_a = 140$ MPa, $N_f = [6.81 \pm 6.53]$ E4

$\sigma_a = 120$ MPa, $N_f = [2.11 \pm 2.19]$ E5

$\sigma_a = 100$ MPa, $N_f = [3.13 \pm 1.76]$ E5

$\sigma_a = 720 - 119.56 \, N_f$

No. of cycles to failure, N_f

Figure 3.37 Fatigue strength of AlSi10Mg in the early high cycle fatigue regime at 20 Hz of testing frequency

Tests on tiny volumes (characteristic of hourglass specimens) would not allow a reliable assessment of the distribution of defect size and the HCF response. Heat-treated AlSi10Mg has been shown that it can be more fatigue resistant than non-heated material. However, a study showed how a small improvement in heat treatment could inflict fatigue resistance in the presence of critical defects [259]. The fractographic analysis in Figure 3.38 supports the origins of failure found in the literature. It is shown how a specimen of AlSi10Mg would fail at clusters of subsurface defects of varying size under a stress amplitude of 120 MPa. A slightly higher than one-third of the specimen fracture area had a stable crack propagation region. The remaining consisted of unstable crack growth and residual fracture. The observation suggests that AlSi10Mg would exhibit a rather fatigue life dominated by crack propagation since crack propagation begins at an earlier stage of fatigue life and continues until complete failure of the specimen. The size of the defect responsible for the fatigue failure can be assessed using Kitagawa-type diagrams as reported in the literature [260]. Intrinsic size influence could also be quantified via Kitagawa type diagrams. In the presence of flaws, matrix improvement by a T6 heat treatment that relieves residual stresses had no effect on the tested fatigue limit in the literature [261].

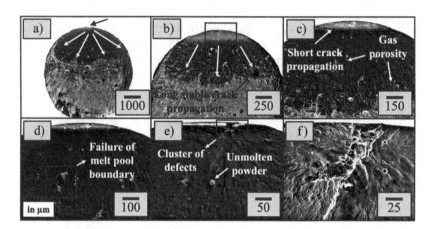

Figure 3.38 Fractographic failure analysis of AlSi10Mg failing at 120 MPa of stress amplitude in a 20 Hz test under scanning electron microscope

In Figure 3.39, the *S-N* relation is shown for Ti-6Al-4V. In the early to the middle of the HCF regime, the fatigue strength suffers from high scatter even when it is relatively less than the scatter in AlSi10Mg. The defects of Ti-6Al-4V had less scatter in terms of morphology, at least in comparison to AlSi10Mg. In addition, the author thinks that microstructural heterogeneity in AlSi10Mg is higher. As shown in the previous section, the Ti-6Al-4V microstructure consists mainly of the prior β grains in the build direction with $\alpha + \beta$ colonies across a wide range of microstructural distances. Thus, as far as the author is concerned, the microstructure of AlSi10Mg is characterized by high variability of melt pool sizes in addition to significantly coarser eutectic morphology at the melt pool boundaries. However, the directional dependency of prior β grains on thermal gradients caused a directionality in fatigue crack growth properties as reported in the literature. Reports in the literature show that stress intensity factor range threshold ΔK_{th} for horizontal specimens is 11.30 MPa·$\sqrt{m}$ while vertical specimens' is 10.30 MPa·$\sqrt{m}$ [262]. As shown in Figure 3.40, porosity plays a critical role in the definition of fatigue strength characteristics of Ti-6Al-4V. Here we observe a failure at 800 MPa of stress amplitude from a pore that is ~250 μm deep from the surface of the specimen. The defect initiated a short crack propagation phase for ~300 μm that is characterized by strain localization and shear delamination in the immediate prior β grain. After that, a more brittle than ductile fracture character is apparent, which predominates crack propagation for a

little less than 1000 μm until the fracture growth becomes more rapid and the
rate of the crack area increases rapidly until the final fracture.

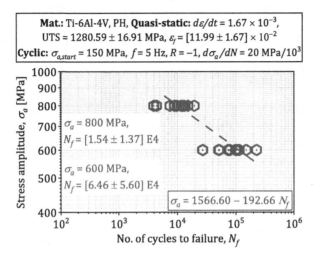

Figure 3.39 Fatigue strength of Ti-6Al-4V in the early high cycle fatigue regime at 5 Hz
of testing frequency

Very high cycle fatigue (VHCF) properties
For characterization of very high cycle fatigue properties, an ultrasonic fatigue
testing system was utilized, which works on the physical principle of resonance
and wave propagation. Vibrations are classified as follows: free vibrations, in
which the vibration is aroused just once and then continue without additional
external impact; damped vibrations, in which friction is pre-present; and mixed
vibrations. An external force that is periodic in nature drives a damped oscillator;
the vibration does not dissipate since the external excitation delivers energy to
the oscillator on a permanent basis [263].

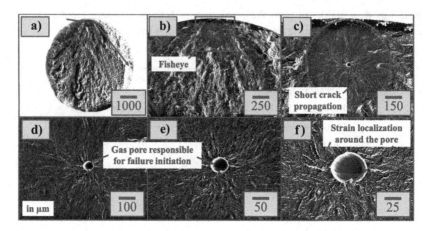

Figure 3.40 Fractographic failure analysis of Ti-6Al-4V failing at 800 MPa of stress amplitude in a 5 Hz test under scanning electron microscope

That is the principle on which the USF-2000A system of Shimadzu is working, the forced vibration concept. However, the vibration between the specimen, the horn, and the booster (see the literature [191] for further illustration of the machine setup and components) is harmonic according to the relation

$$u(t) = A\cos(\omega t + \phi) \tag{3.11}$$

where u is the displacement, t is the time, A is the excitation amplitude, ω is the angular frequency, and ϕ is the phase shift. Since ultrasonic vibration in metallic materials generates temperature because of the internal resistance to cyclic deformation, a pulse:pause scheme, in addition to the application of compressed dry air, is applied to the fatigue specimen in the ultrasonic setting. Therefore, in the pausing phase, the type of vibration is changed to damped vibration. That theoretically means that the specimen comes to rest after a long enough time of lost energy due to friction. The energy of this system can be classified into kinematic $E_{kin}(t)$ and potential $E_{pot}(t)$ energies according to the relation

$$E_{kin}(t) + E_{pot}(t) = \frac{mA^2\omega^2}{2} = \frac{k_sA^2}{2} = constant \tag{3.12}$$

where m is the mass, and k_s is the stiffness of the system. Since the testing system is a multi-component forced vibration system, the natural frequency resonance is essential. For the requested specimen side amplitude by the user to be accurately applied, the natural frequency excitation by the piezoelectric actuator is a prerequisite. In this case, the accurate stress state requested by the user is ensured to occur at the gauge section of the specimen. Thus, the vibration case that depends on the natural frequency gets reduced then from a forced vibration into a free ultrasonic vibration that resonates in the range of ultrasonic frequencies of 20 kHz according to [263].

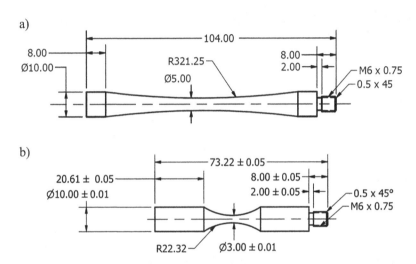

Figure 3.41 Technical drawing of specimen geometry of the ultrasonic fatigue test at 20 kHz: a) AlSi10Mg; b) Ti-6Al-4V

Since VHCF testing occurs in the range of stresses where strains are predominantly elastic and plastic strains are considered negligible [256, 264], only the elastic properties of the specimens contribute a role to the natural frequency resonance of the system consisting of the specimen, horn, and booster. The elastic moduli based on which the specimen geometries of Figure 3.41 for AlSi10Mg and Ti-6Al-4V were designed can be found in Figure 3.27 and Figure 3.28. To control the accumulation of temperature at ultrasonic frequencies, a pulse:pause ratio of a duration of 200 ms was applied on a 1:1 basis. In addition, active specimen cooling by pressurized dry air was applied. Hence, the temperature increase was kept under 10 K and was monitored using a pyrometer of Calex Electronics Limited Type PU301.

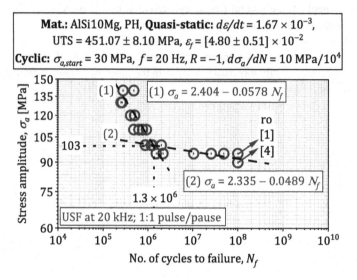

Figure 3.42 Fatigue strength of AlSi10Mg in the late high cycle into very high cycle fatigue regime tested at 20 kHz

In Figure 3.42, the ultrasonic fatigue data of AlSi10Mg is presented from the late HCF regime into the VHCF regime. It was observed that between 140 and 100 MPa, the life does not increase significantly, and the scatter is relatively less in comparison to fatigue strength between 100 and 90 MPa of stress amplitude. At these stress amplitude levels, the fatigue strength is more horizontally distributed with high scatter, and therefore 4 runouts take place at 90 MPa for a limiting number of cycles of 1E8. For Ti-6Al-4V, the life increases rapidly in the range of 650 to 550 MPa of stress amplitude with a low slope S-N relation, according to Figure 3.43. Below 574 MPa of stress amplitude, the life increases less rapidly, although scatter significantly increases from early VHCF until the limiting number of cycles of 1E9. This indicates a significant difference in mechanism transformation with respective fatigue regimes from AlSi10Mg to Ti-6Al-4V. It was mentioned in the previous section how the damage mechanisms in Ti-6Al-4V will transfer to the inner α colonies that have a higher capacity for strain.

Therefore, when the strain rate is high, yet the load is low, the motion of dislocations is not reaching the prior β boundaries, and the strain is localized within the α colonies, which are relatively more strain sensitive. Thus, the damage is more rapid at this *S-N* range, and life increase due to low reduction is not as significant as in earlier higher load experiments. The influence of fatigue testing frequency on the resulting fatigue strength has been debated to a large extent in the literature [265].

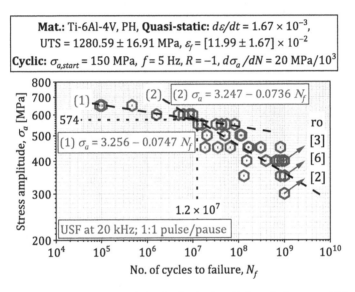

Figure 3.43 Fatigue strength of Ti-6Al-4V in the late high cycle into very high cycle fatigue regime tested at 20 kHz

In Figure 3.44, the influence of frequency is deduced by comparing fatigue data for AlSi10Mg and Ti-6Al-4V, which were tested both at 20 and 5 Hz, respectively, and 20 kHz for each of them. It is evident in both cases that higher frequency led to a higher life. Although in the case of AlSi10Mg, it does not appear to be load-dependent, in Ti-6Al-4V, it is load-dependent. Already in the literature, the frequency effect was apparent in Ti-6Al-4V at high frequencies, which led to an increase in fatigue strength [266]. If crack development rates are less than the mean diffusion distance of hydrogen, humidity affects ultrasonic fatigue crack growth. The *S-N* curve is altered roughly 30%—40% towards lower stress levels while as reported for 2024-T351 in distilled water. Therefore, there should be an influence of the dry air chemical composition on the crack tip propagation in comparison to a low-frequency test that is conducted under ambient conditions [267].

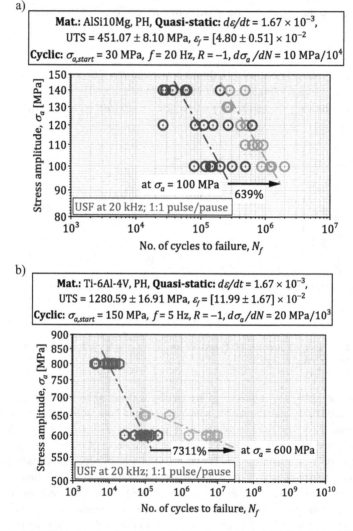

Figure 3.44 Effect of testing frequency of fatigue strength: a) AlSi10Mg; b) Ti-6Al-4V

During the ultrasonic fatigue testing, the test stopping criterion was set to ± 300 Hz to stop the fatigue test before the complete separation of fracture surfaces. This allowed a further μ-CT scanning of the specimen in its entirety and a three-dimensional (3D) visualization of the crack path inside the specimen. There have been reports in the literature discussing 3D fatigue crack morphology using, for instance, digital volume correlation (DVC). A technique for detecting failure commencement has been developed that is based on the relative deformation of the turbine blades during operation. Two digital cameras were mounted in front of the wind turbine to photograph the rotor blade deformation [268]. On the other hand, in μ-CT, the smallest detectable feature at a certain projection magnification depends on the total sharpness around crack flanks, gradually decreasing the image contrast and the system's detectability [188]. In Figure 3.45, the interaction of the X-ray beam with a fatigue crack opening is schematically explained. The distorted attenuation coefficients of the specimen are compensated by the application of higher beam voltage. This was possible with the current system discussed earlier since the investigated alloys here are light metals of AlSi10Mg and Ti-6Al-4V with 2.76 and 4.43 g/cm^3, respectively.

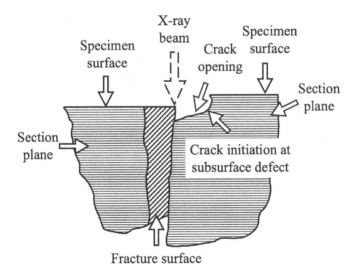

Figure 3.45 Illustration of the scanning of crack openings using X-ray microcomputed tomography

In Figure 3.46, the 3D crack morphology of an AlSi10Mg specimen is shown where the crack initiation location was obvious to happen at a cluster of defects in the subsurface of the specimen. This is well in agreement with the fractographic observation in Figure 3.38. The crack front meats several defects along the propagation front. Since load is very low in the VHCF range, the crack propagation line showed linear elastic characteristics. However, after the crack front crossed half of the specimen, it seems that the specimen compliance becomes so high, and stiffness is so degraded such that the crack front starts to plasticize and deflect possibly in the direction of the lowest resistance. On the other hand, another possibility for this deflection would be that the crack front has faced some microstructural discontinuities, which exist in plenty of quantity in the vertical direction in AlSi10Mg.

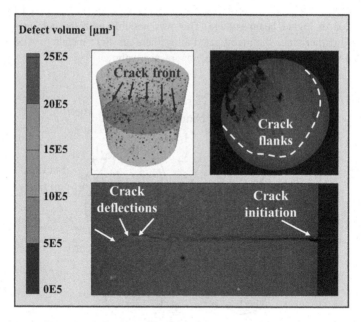

Figure 3.46 Three-dimensional very high-cycle fatigue crack analysis of AlSi10Mg at 95 MPa of stress amplitude

On the contrary, in Figure 3.47, the crack initiation in Ti-6Al-4V appears at a specific defect in the subsurface, which is much smaller than the defect cluster that caused failure in AlSi10Mg. However, failure proceeds in both directions to the surface and in the bulk of the specimen following an inclination indicating a shear delamination mechanism. This Ti-6Al-4V specimen contained significantly less residual porosity than AlSi10Mg; therefore, the crack front meeting further pores was not so evident in Ti-6Al-4V. At the same time, the crack front is obviously nonplanar and follows local ductile fracture damage due to the colonies of the α phase, which have inclination angles that are highly scattered in orientation. Towards the end life of the specimen, the crack front deflects sharply, giving an indication of highly anticipated plastic deformation in the residual fracture zone.

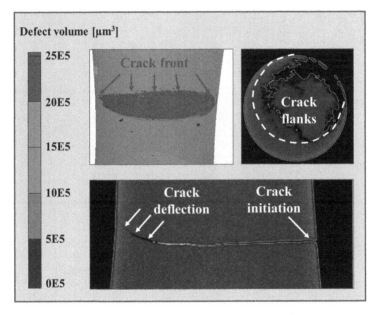

Figure 3.47 Three-dimensional very high-cycle fatigue crack analysis of Ti-6Al-4V at 400 MPa of stress amplitude

Figure 6.7:

Estimation of Lifetime Trends Based on FEM

4

4.1 Linear Elastic Fracture Mechanics

Linear elastic fracture mechanics (LEFM) is an area of mechanics that performs mechanical analysis on fracture surfaces using the linear elastic theory. Some fatigue fracture criteria factors, such as the stress intensity factor K and the energy release rate G, are based on LEFM. The material is assumed to be isotropic in LEFM, and only linear elastic deformation is examined. Plastic deformation and anisotropic materials are not considered in LEFM. As a result, only small-scale yielding is allowed. The theory of elastic-plastic fracture mechanics (EPFM) should be used for large-scale yielding [269].

Figure 4.1 Principal modes of crack formation and propagation [270]

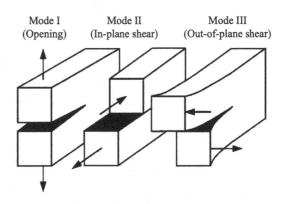

Mode I (Opening) Mode II (In-plane shear) Mode III (Out-of-plane shear)

Content in this chapter is partially based on publication [304]

M. Mamduh Mustafa Awd, *Machine Learning Algorithm for Fatigue Fields in Additive Manufacturing*, Werkstofftechnische Berichte | Reports of Materials Science and Engineering, https:doi.org/10.1007/978-3-658-40237-2_4

There are three loading modes for fracture development [270]; see Figure 4.1. The loading path for mode I is perpendicular to the fracture plane [269]. The in-plane shear mode is mode II. Shear stresses run in the same direction as the crack plane. The out-of-plane shear mode is the third mode. One crack plane slides on top of the other under shear loads, and the crack front is parallel to the sliding direction [269]. In real-world settings, a mixed mode, which is a mixture of three basic modes, is frequent. Mode I is the most hazardous.

4.1.1   Description of Crack Tip State

It is supposed that a penetrating fracture runs through the infinite plate's center. As illustrated in Figure 4.2, the crack length is *2a* of a penny-shaped crack in an infinite plate, and the section thickness is *B*. The loading vector is perpendicular to the fracture plane, and uniform tensile distant stresses are applied on both edges [269].

Figure 4.2 A finite-field crack in an infinite plate with remote loading with stress state at a crack tip [269]

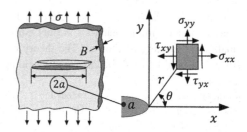

Irwin succeeded in describing the element's stress fields at the crack tip in 1957 [270]. Planar stresses, such as normal and shear loads, are applied to the element in Figure 4.2. Equation 4.1, Equation 4.2, and Equation 4.3 [271] can be used to represent these stress components.

$$\sigma_{xx} = \frac{K_I}{\sqrt{2\pi r}} \cos\left(\frac{\theta}{2}\right)\left[1 - sin\left(\frac{\theta}{2}\right)sin\left(\frac{3\theta}{2}\right)\right] \tag{4.1}$$

$$\sigma_{yy} = \frac{K_I}{\sqrt{2\pi r}} \cos\left(\frac{\theta}{2}\right)\left[1 + sin\left(\frac{\theta}{2}\right)sin\left(\frac{3\theta}{2}\right)\right] \tag{4.2}$$

$$\tau_{xy} = \frac{K_I}{\sqrt{2\pi r}} \cos\left(\frac{\theta}{2}\right)sin\left(\frac{\theta}{2}\right)cos\left(\frac{3\theta}{2}\right) \tag{4.3}$$

θ and r are illustrated in Figure 4.2. From the equations, the stress components are proportional to $1/\sqrt{2\pi r}$. Regardless of the shape or loading, at $r = 0$, the stress components approach infinity. The stress singularity [270] is the name for this phenomenon. Even while infinite loads can cause components near the crack tip to shatter in theory, the crack propagates in fact only when external loadings surpass the fatigue strength threshold. Because when the tension surpasses the yield point, plastic deformation occurs. As a result, using stress fields to forecast fractures is absurd. Consequently, the stress intensity factor (SIF) is used to depict the fracture tip's stress intensity. Remote stresses, fracture size, and crack form all play a role [269]. The SIF for mode I is K_I. K_I rises as the number of distant stressors and fracture dimensions grow. When K_I reaches the crucial value K_{Ic}, the fracture starts to spread uncontrollably [269]. K_{IC} stands for fracture toughness and is a material characteristic that measures a material's resistance to fracture. K_{IC} has no relation to the part's form and geometry [269]. Hellen discovered a novel fracture criterion on the energy level in 1956 [272]. This criterion has the benefit of being simple to use in actual situations. When enough energy is given to overcome the material's resistance, the fracture expands. G is defined by Irwin as the amount of energy required to create a fracture [269], as illustrated

$$G = -\frac{d\Pi}{dA} \tag{4.4}$$

A is the area of the crack, and Π is the total potential energy [271]. It shows how quickly energy is released as the fracture spreads. In Figure 4.2, G can be expressed as [274]

$$G = \frac{\pi \sigma^2 a}{E} \tag{4.5}$$

G_c is the critical energy release rate and when G reaches G_c, the crack propagates rapidly [269]

$$G_c = \frac{dW_s}{dA} = 2w_f \tag{4.6}$$

W_s is the work needed to create the crack surfaces and w_f is the fracture energy [269]. Both K and G are critical variables that can be utilized to forecast crack growth. For mode I, G_I and K_I can be expressed by [269]

$$G_I = \frac{K_I^2}{E'} \tag{4.7}$$

For plane strain $E' = E/(1 - v^2)$, for plane stress $E' = E$ [269], where E is elastic modulus and v is Poisson's ratio. The relationship between G_{Ic} and K_{Ic} is shown in Equation 4.8 [269].

$$G_{Ic} = \frac{K_{Ic}^2}{E'} \tag{4.8}$$

Rice first proposed the J-Integral idea in 1968 [273]. It is compatible with both LEFM and EPFM. The J-Integral quantifies the energy produced from the body as a crack propagates. The J-Integral route circles the crack front in a clockwise direction. Starting and ending points of contour, as illustrated in Figure 4.3, are arbitrary positions on the bottom and upper crack surfaces, respectively [274].

Figure 4.3 Arbitrary contour $\varGamma$ of a J-integral [270]

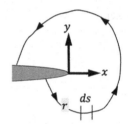

The two-dimensional J-integral can be expressed by [270]

$$J = \int\limits_\varGamma \left(w\,dy - T_i \frac{\partial u_i}{\partial x}\,ds \right) \tag{4.9}$$

w is the strain energy density, T_i is the stress vector, u_i is the displacement vector and ds is the increment along the path $\varGamma$ [270]. In [271], two arbitrary contours $\varGamma_1$ and $\varGamma_3$ with opposing rotating orientations are shown. $\varGamma_2$ and $\varGamma_4$ combine them together and form a closed contour $\varGamma$. Therefore $J_\varGamma = J_{\varGamma_1} + J_{\varGamma_2} + J_{\varGamma_3} + J_{\varGamma_4} = 0$ [269]. On the crack fragment, $T_i = dy = 0$, so $J_{\varGamma_2} = J_{\varGamma_4} = 0$ [270]. Therefore $J_{\varGamma_1} = -J_{\varGamma_3}$. It can be deduced that any arbitrary counterclockwise contour $\varGamma$ has the same J-Integral value. The J-Integral is path unbiased [270]. With linear elastic materials, J-integral in a mixed-mode can be stated by G and K [270]

$$J = G = \frac{K_I^2}{E'} + \frac{K_{II}^2}{E'} + \frac{1+\upsilon}{E'}K_{III}^2 \qquad (4.10)$$

then the J value reaches a critical value J_{Ic}, the crack propagates rapidly. As revealed [270], J_{Ic} can be addressed by G_{Ic} and K_{Ic} [270]

$$J_{Ic} = G_{Ic} = K_{Ic}^2 \left(\frac{1}{E'}\right) \qquad (4.11)$$

Crack growth

Maximum tangential stress criterion (MTS), Maximum energy release rate (MERR) criterion, and $K_{II} = 0$ criterion; are three techniques in Abaqus for predicting the crack propagation direction [275]. If the greatest tangential stress approaches the critical value, fractures will spread in the direction perpendicular to it, according to MTS [276]. Erdogan and Sih [277] assert that

$$\hat{\theta} = cos^{-1}\left(\frac{3K_{II}^2 + \sqrt{K_I^4 + 8K_I^2 K_{II}^2}}{K_I^2 + 9K_{II}^2}\right) \qquad (4.12)$$

$\hat{\theta}$ is the angle in comparison to the first fracture plane [275]. For mode I, $K_{II} = 0$. Therefore $\hat{\theta} = 0$ and the crack will grow in the same direction it originated [275]. MTS is frequently used since it is easy to use and has a basic form. At the crack front, energy is released when the crack propagates. According to the first law of thermodynamics, when G_{max} reaches G_c. Fractures spread in the direction that can release the most energy [275]. The crack will expand in the same direction it originated as $K_{II} = 0$. Only assuming homogenous and isotropic materials is suitable for the application of this theory [278].

4.1.2 Fatigue Crack Propagation

The fatigue crack growth rate can be used to determine overall cycle numbers as well as to anticipate how cracks grow [279]. The fatigue crack growth rate is a measurement of how quickly a crack expands when subjected to fatigue loading. The value of crack growth per cycle is $\Delta a/\Delta N$ if the crack propagates for an Δa in ΔN cycles. The fatigue crack propagation rate can be expressed as da/dN as the time interval approaches zero. The power-law relationship between da/dN and the stress intensity factor changes K is shown in Equation 4.13, which indicates that da/dN and the stress intensity factor changes K have a power-law relationship [269]

$$\frac{da}{dN} = C(\Delta K)^m \tag{4.13}$$

parameters C and m are material constants [269]. ΔK is the difference between maximum and minimum stress intensity factors [270]. Linearizing Equation 4.13 with the logarithmic scale, we obtain

$$log\frac{da}{dN} = \log C + mlog(\Delta K) \tag{4.14}$$

In [270, 271], the curve between ΔK and da/dN is plotted on a $log - log$ scale. This curve is divided into three parts. The fracture will not develop if ΔK is less than the threshold stress intensity factor ΔK_{th} [280].

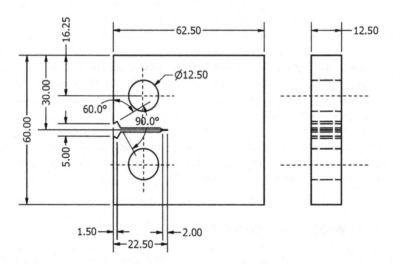

Figure 4.4 Technical drawing of a compact tension specimen according to ASTM E647 [281]

When ΔK reaches K_c, da/dN becomes rapid, and the fracture propagates in an unsteady manner [282]. Only da/dN and K in region II adhere to the Paris law concept [256]. A notched sample is used for the compact tension specimen. When cyclic loadings are applied to it, fracture begins at the notch point and spreads [281]. Figure 4.4 shows the geometry of a compact tension (CT) specimen [281]. The width W is 50 mm, the thickness B is 12.5 mm, and a is the crack length. For CT-specimen ΔK is calculated as follows [281]. ΔF is the applied force range on the specimen, $\alpha = a/W$, $\Delta K = K_{max} - K_{min}$ and K_{max} is maximum SIF which corresponds to F_{max}. K_{min}: when $R > 0$, the minimum SIF K_{min} corresponds to F_{min} and when $R \leq 0$, $K_{min} = 0$. Equation 4.15 explains the relationship between these parameters.

$$\Delta K = \frac{\Delta F}{B\sqrt{W}} \frac{(2+\alpha)}{(1-\alpha)^{\frac{3}{2}}} \left(0.886 + 4.640\alpha - 13.320\alpha^2 + 14.720\alpha^3 - 5.600\alpha^4\right)$$

$$(4.15)$$

4.2 Extended Finite Element Method

The continuous function is utilized as the interpolation function in the generalized finite element method (GFEM). The form function must be continuous, and the material characteristic must not change abruptly [283, 284]. When modeling the crack propagation process using GFEM, the mesh density must be very high, and remeshing is required after every single increment to achieve an accurate result [285, 286]. For GFEM, the definition of the displacement field is related to the nodes of elements. The displacement vector function can be expressed by the nodal shape function $N_I(X)$ and the displacement of nodes $\vec{u}_I$, as shown in [285]

$$\vec{u} = \sum_{I=1}^{N} N_I(x)\vec{u}_I \qquad (4.16)$$

x is coordinate, and I is the element node label. After adding enrichment functions because of Equation 4.16, the displacement vector function of the extended finite element method (XFEM) can be written as

$$\vec{u} = \sum_{I=1}^{N} N_I(x) \left[\vec{u}_I + H(x)\vec{a}_I + \sum_{\alpha=1}^{4} F_\alpha(x) \vec{b}_I^\alpha \right] \qquad (4.17)$$

$N_I(x)$ is the nodal shape function, $\vec{u}_I$ is the displacement of nodes, $H(x)$ is the heavy side function, $\vec{a}_I$ nodal enriched degree of freedom (DOF), $F_\alpha(x)$ is the crack tip asymptotic function, $\vec{b}_I^\alpha$ the nodal DOF. In Equation 4.17, all nodes have access to the first component.

Figure 4.5 Jump function of the polar coordinates [285]

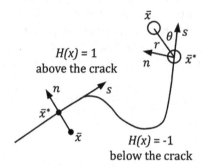

The second part is the Heaviside enrichment term, which can be used to characterize the discontinuous displacement field for components that are completely separated by a fracture [287]. Singularity is dealt with by the third component [288]. $H(x)$ is the Heaviside distribution-following jump enrichment function, as shown in Equation 4.18. $\vec{a}_I$ is nodal enriched DOF for jump discontinuity [283]

$$H(x) = \begin{cases} 1 & if(\vec{x} - \vec{x}^*)\vec{n} \geq 0, \\ -1 & otherwise \end{cases} \tag{4.18}$$

$\vec{x}$ is the integration point, $\vec{n}$ is the unit vector that is vertical to the crack surface, and $\vec{x}^*$ is the intersection point of $\vec{n}$ and the crack surface [283, 285]. These parameters are shown in Figure 4.5. $F_\alpha(x)$ in Equation 4.17 is the asymptotic function that accounts for the fracture tip singularity. It is only applicable for linear elastic materials that are isotropic [289]. It can be expressed in Equation 4.19 [281]. $\vec{b}_I^\alpha$ is nodal DOF for crack tip enrichment [286].

$$[F_\alpha(x)] = \left[\sqrt{r}sin\frac{\theta}{2}, \sqrt{r}cos\frac{\theta}{2}, \sqrt{r}sin\theta sin\frac{\theta}{2}, \sqrt{r}sin\theta cos\frac{\theta}{2}\right] \tag{4.19}$$

r and θ are polar coordinate values. The leap function is used on nodes with circles, while the crack tip function is used on nodes with squares in Figure 4.6.

The blue element represents the crack tip, the yellow elements are enriched elements, and the pink elements are blending elements that include both enriched and regular nodes [285].

Figure 4.6 Illustration of enriching nodes within the radius of an extended finite element method field in a two-dimensional model [285]

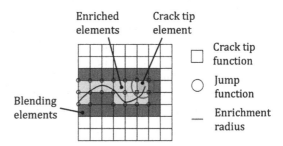

XFEM does not have a high demand on the mesh density, and when simulating the crack propagation process, remeshing is not needed [290]. However, the region at which the crack propagates should have a refined mesh to obtain an accurate SIF value. This will occupy a high computational power [285]. Another drawback is that Abaqus only provides static SIF computation. SIF outputs for a propagating crack are not possible [291].

4.2.1   SIF Analysis Using XFEM and Contour Integral Method

Stress intensity factor (SIF) values in static steps can be directly outputted in Abaqus. XFEM and the contour integral method are the two forms of crack definitions available in Abaqus. These two approaches are utilized in this section to create models with varying crack lengths [286]. The Python script is used to automatically modify the fracture length from 2 mm to 38 mm to test the accuracy of SIF with varied crack lengths. The analytical results will be compared to the SIF values derived by Abaqus. To record the J-Integral and SIF values, two history outputs named H-Output-J-Integral and H-Output-SIF are established. The contour integral approach is used by Abaqus to calculate these values. Only the last result is outputted since the domain is Crack-1 and the frequency is the last increment [286]. There are 5 contours in all. To deal with the singularity problem, a circle with a diameter of 2 mm is utilized to ring the crack point. The contour integral technique cannot be used if the instance is dependent. The crack front is specified in the q vector $(1,0,0)$ and defines which direction the crack propagates in [292]. Two reference points are generated before couplings are made to represent pins hitting holes [293]. There is no relative displacement between the reference point and all slave nodes because of the kinematic coupling [274]. The reference point and slave nodes can move relative to each other to distribute coupling. As a result, distributing coupling has a lower rigidity than kinematic coupling [293]. The crack tip, not the stresses around holes, is the focal point of this simulation. As a result, the kinematic coupling is used. In this simulation, an AlSi10Mg alloy with an elastic modulus of ~73 GPa and a Poisson's ratio of 0.3 is used. The applied force range ΔF is 2 kN. The CT specimen is divided into 12 pieces to ensure that the mesh is spread evenly. The distance between every two global seeds is 1 mm, which is the size of the global seeds. The break-in of this model is in the central plane of the CT specimen. As a result, to improve accuracy, the center area should be finer meshed. Local seeds are generated in the center area, and the element size is 0.5 mm. C3D8R is the element type, and the shape is linear hexahedron [292].

The K_I values obtained using XFEM and the contour integral technique can be compared to analytical values. The analytical SIF values are computed using Equation 4.15 [281], which is valid for CT-specimen. Figure 4.7 shows the relationship between fracture lengths and K_I values. One approach is represented by each curve. The general tendencies of these three techniques can be determined to be similar. Equation 4.20 is used to determine the inaccuracy of XFEM K_I based on analytical K_I. Contour integral K_I error is determined based on Equation 4.21.

$$XFEM\,error = \frac{XFEM\,K_I - Theoretical\,K_I}{Theoretical\,K_I} \qquad (4.20)$$

$$Contour\,integral\,error = \frac{Contour\,integral\,K_I - Theoretical\,K_I}{Theoretical\,K_I} \qquad (4.21)$$

As seen in Figure 4.8, the XFEM method's inaccuracies decrease as the crack develops.

At first, the accuracy of the contour integral technique improves. The accuracy of the contour integral technique drops dramatically when the fracture length exceeds 43 mm. The contour integral technique places a great demand on mesh density, which might be the explanation. When the fracture length grows excessively long, the stress intensity skyrockets. The precision is harmed by the large deformation of the mesh. Because XFEM is a mesh-independent technique, it does not have a high mesh density need. When the fracture length increases, the error decreases. Overall, the contour integral technique outperforms XFEM in

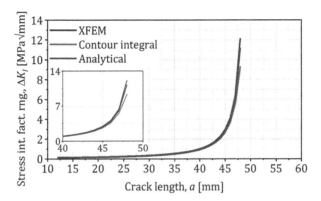

Figure 4.7 Comparison of mode I stress intensity factor range K_I in extended finite element method, contour integral and analytical methods for AlSi10Mg at $R = 0.1$

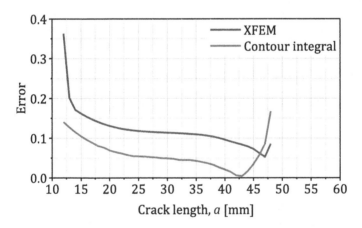

Figure 4.8 Illustration of relative error between extended finite element method and contour integral methods in the standard crack propagation simulation on compact tension specimen for AlSi10Mg at $R = 0.1$

terms of accuracy. Because it can help model any fracture route, the extended finite element approach is used. The fatigue crack growth mechanism is modeled in the next section using XFEM. After simulation, the link between fracture lengths and cycle numbers can well be determined.

4.2.2 Fatigue Crack Propagation with XFEM

The procedure type is direct cyclic, and a new step named Step-1 is generated. The cycle time is set to the time period in the initial tab, and the nonlinear geometry effects (Nlgeom) are turned off because the study of nonlinear geometric behavior under direct cyclic step is not supported by Abaqus. The maximum number of increments in a single loading cycle is 1000 [293]. The analysis will be terminated if the practical increment number surpasses 1000. Because the modeling of fatigue crack development is complex, the value is set to 1000 instead of the default value of 100. The global Newton iterations are utilized to modify the displacement Fourier coefficients in the direct cyclic step. Abaqus/CAE uses an adaptive method for determining Fourier terms [292]. Abaqus starts with a small number of Fourier terms and checks if they satisfy the criterion. The number of Fourier terms will be raised to meet the equilibrium if the condition is not met. The starting and the maximum number of Fourier terms are chosen to be 10 and 25, respectively, because additional Fourier terms can improve computational accuracy [294]. It should be noted that having too many Fourier terms might take up a lot of computational resources. Because the damage extrapolation approach is used in the direct cyclic stage, the minimum and maximum cycle increment sizes must be determined. This method is employed to accelerate the convergence process [295]. Suppose the initialization requirement is met at cycle N, the new damage variable $D_{N+\Delta N}$ for the following increment is computed, as illustrated in Equation 4.22. If $D_{N+\Delta N}$ satisfies the condition, the fracture will propagate forward by N cycles.

$$D_{N+\Delta N} = D_N + \frac{\Delta N}{L_c} c_3 \Delta w^{c_4} \qquad (4.22)$$

L_c is the characteristic length and c_3 and c_4 are parameters for crack growth [296]. Because the cycle numbers in fatigue loading tests are so large, the minimum and maximum cycle increment sizes have been set to 1E2 and 1E3, respectively, to assist in speeding up the computation. Damage extrapolation tolerance is set to 1, which aids in damage extrapolation accuracy control [292, 293]. The signed distance function PHILSM, PSILSM, STATUSXFEM, and STATUS are field output variables. To increase the efficiency for severely discontinuous behavior, the discontinuous analysis at the general solution controls tab for Step-1 is toggled on the frequency of checking the iteration residuals $I_0 = 8$ and staring point of the convergence checks $I_R = 10$ [293]. I_A is set to 50 in the time incrementation, which is the maximum number of cutbacks permitted.

Figure 4.9   Failure of elements in the enriched extended finite element method region along the crack path with STATUS XFEM shown

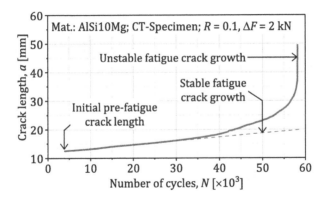

Figure 4.10   Crack length evolution as the simulation continues and the number of cycles increases indicating the stable and unstable fracture

The convergence of the computation can be improved by increasing I_A [297]. The virtual crack closure technique (VCCT) or improved VCCT fracture criterion can only be created in Abaqus graphical user interface (GUI). The fatigue fracture criterion should be introduced by modifying keywords. The fracture will not propagate unless the fatigue fracture criterion is included in the keywords [298]. The CT specimen is divided into 14 parts initially. Global seeds are 0.5 mm in size. CPS4R is used since it is 4-node bilinear plane stress reduced integration with hourglass control [274].

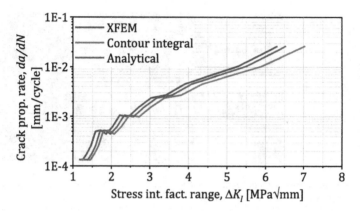

Figure 4.11 Paris region of ΔK vs. da/dN according to an analytical model, extended finite element method and contour integral methods

Three parts are assigned to the quad element type, free technique, and the option to utilize mapped meshing when applicable is toggled on in the mesh settings. The plot of STATUSXFEM is given in Figure 4.9. These red elements have a STATUSXFEM value of 1.0, indicating that they have completely failed. The values of crack lengths and cycle counts can be retrieved once Abaqus has completed the fatigue crack propagation simulation. Figure 4.10 depicts the data and plots the curve. One pair of data is taken from every five pairs of data to decrease stochastic error and achieve a more stable outcome. Equation 4.23 and Equation 4.24 are used to calculate the values of $(da/dN)^{\overline{a}}$ and $\overline{a}$ [299].

$$\left(\frac{da}{dN}\right)_{\overline{a}} = \frac{a_{i+1} - a_i}{N_{i+1} - N_i} \tag{4.23}$$

$$\overline{a} = \frac{a_{i+1} + a_i}{2} \tag{4.24}$$

The SIF can be calculated using three different methods: XFEM, contour integral, and analytical. The Python script is used to generate XFEM and the contour integral cracks iteratively. Three groups of ΔK_I values are produced using the XFEM, contour integral, and analytical methods. Because the focus of this analysis is on Region II, which follows the Paris law, the data in Region III has been removed. da/dN is shown as a function of ΔK_I on a logarithmic scale in

Figure 4.11 [300]. The analytical technique and XFEM have good fitting accuracy for parameter m, while the analytical approach has the greatest accuracy for the parameter $\log C$ when compared to XFEM and contour integral method. Both in terms of fitting parameter $\log C$ and m, the contour integral approach has the lowest accuracy according to Table 4.1.

Table 4.1 Errors of $\log C$ and m based on the experimental values

	Analytical method	XFEM method	Contour integral method
Simulation m	2.97	2.96	3.08
Experimental m	2.81	2.81	2.81
Error m	5.70%	5.59%	9.90%
Simulation $\log C$	−5.80	−5.95	−6.20
Experimental $\log C$	−5.78	−5.78	−5.78
Error $\log C$	0.35%	2.94%	7.27%

4.3 Visualization and Post-processing

The fatigue crack growth rate da/dN represents how quickly the crack propagates; therefore, it can be used to forecast durability [286]. The change in stress intensity factor range (ΔK) or the energy release rate range (ΔG) is represented on the x-axis of the curve in Figure 4.12 where G is the energy release rate. The curve in Figure 4.12 can be split into three sections where the fracture will not develop if ΔG is less than ΔG_{th} [276]. If ΔG is more than G_{pl} the fracture will spread at a faster pace and become unstable [277]. By default, $G_{pl} = 0.85 \cdot G_C$ and G_{pl} is the energy release rate upper limit. G_C is the fracture energy release rate. The connection between ΔG and da/dN follows the Paris law in the middle region if $\Delta G_{th} < \Delta G < G_{pl}$ as shown [278]

$$\frac{da}{dN} = c_3(\Delta G)^{c_4} \qquad (4.25)$$

If a log-log scale is applied to Equation 4.25, the region II curve is linear. The slope is c_4 and the intercept is $\log(c_3)$

$$\log\left(\frac{da}{dN}\right) = \log(c_3) + c_4 log(\Delta G) \qquad (4.26)$$

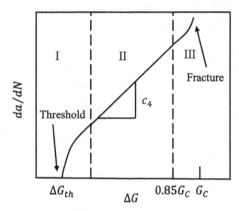

Figure 4.12 Fatigue crack growth rate as a function of the energy release rate following the Paris law [271]

4.3.1 Virtual Crack Closure Technique (VCCT)

The virtual crack closure technique (VCCT) is a popular method for calculating energy release rates that can be used to mimic crack propagation in a simulation by debonding the components over a pre-defined surface. The essential concept of VCCT for calculating G is that the energy produced during crack propagation equals the energy required to seal the crack [273]. Furthermore, if the element size is small enough, the opening displacements for the virtual crack tip $(\Delta a + a)$ and the real crack tip (a) is about identical. G_I can be obtained by

$$G_I = \frac{v_{1,6}F_{y,2,5}}{2bd} \qquad (4.27)$$

$v_{1,6}$ is the displacement between node 1 and node 6 in the y-direction. $F_{y,2,5}$ is the force between node 2 and node 5 in the y-direction. d is the element length, and b is the width. For the opening mode, when $G_I > G_{IC}$, node 2 and node 5 will begin to release. G_{II} and G_{III} are the energy release rates for mode II and mode III and can be calculated as [273]

$$G_{II} = \frac{u_{1,6} F_{x,2,5}}{2bd} \tag{4.28}$$

$u_{1,6}$ is the displacement between node 1 and node 6 in the x-direction. $F_{x,2,5}$ is the force between node 2 and node 5 in the x-direction

$$G_{III} = \frac{w_{1,6} F_{z,2,5}}{2bd} \tag{4.29}$$

$w_{1,6}$ is the displacement between node 1 and node 6 in the Z direction. $F_{z,2,5}$ is the force between node 2 and node 5 in the z-direction. For a mixed-mode, whether the crack will propagate is determined by the fracture criterion f. G_{equiv} and G_{equivc} are the equivalent energy release rate and critical equivalent energy release rate and $f = G_{equiv}/G_{equivc}$. When $1.0 \leq f \leq 1.0 + f_{tol}$, the crack propagates. Here, f_{tol} is the tolerance. There are three methods to calculate G_{equiv} and G_{equivc}, which are the power law, Reeder law, and BK law [274]. G_{Ic}, G_{IIc} and G_{IIIc} are critical energy release rates for three modes. a_m, a_n and a_0 are exponents for Power law. η is the exponent for BK law and Reeder law. The following is the calculation method

$$\frac{G_{equiv}}{G_{equivc}} = \left(\frac{G_I}{G_{Ic}}\right)^{a_m} + \left(\frac{G_{II}}{G_{IIc}}\right)^{a_n} + \left(\frac{G_{III}}{G_{IIIc}}\right)^{a_0} \tag{4.30}$$

$$G_{equivc} = G_{Ic} + (G_{IIc} - G_{Ic})\left(\frac{G_{II} + G_{III}}{G_I + G_{II} + G_{III}}\right)^{\eta} \tag{4.31}$$

$$G_{equivc} = G_{Ic} + (G_{IIc} - G_{Ic})\left(\frac{G_{II} + G_{III}}{G_I + G_{II} + G_{III}}\right)^{\eta}$$

$$+ (G_{IIIc}$$

$$- G_{IIc})\left(\frac{G_{III}}{G_{II} + G_{III}}\right)\left(\frac{G_{II} + G_{III}}{G_I + G_{II} + G_{III}}\right)^{\eta}$$

$$= G_{equiv} = G_I + G_{II} + G_{III} \tag{4.32}$$

The G_{Ic}, G_{IIc} and η parameters are the only ones necessary for BK law. G_{Ic}, G_{IIc}, G_{IIIc} and three exponents a_m, a_n, and a_0 are necessary for Power law. The Reeder law is derived from the BK law with the inclusion of the G_{IIIc} component. The Reeder law is used on the assumption that G_{IIIc} is not equivalent to G_{IIc}. The Reeder law is the same as the BK law if $G_{IIc} = G_{IIIc}$. Only a three-dimensional model can be utilized with the Reeder law [274]. Because the exponents can alter the proportions of the components in three directions, the Power-law is the most versatile. As a result, it is extensively used and will be applied in this investigation.

4.3.2   Level Set Method

The level set method (LSM) is used to locate the fracture front [280]. The smallest distance between the node and the crack surface is represented in Figure 4.13 as $\underset{x_\Gamma \in \Gamma(t)}{\min} \|x - x_\Gamma\|$.

Figure 4.13 Schematic principle of the level set method to locate a crack in Abaqus [301]

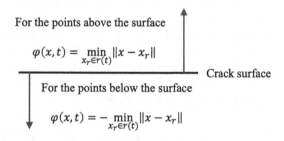

For the points above the surface

$$\varphi(x, t) = \min_{x_r \in r(t)} \|x - x_r\|$$

Crack surface

For the points below the surface

$$\varphi(x, t) = -\min_{x_r \in r(t)} \|x - x_r\|$$

$\varphi(x, t)$ is positive if the node is above the surface and negative if the node is below the surface. It's simple to tell which side the nodes are on when you use the LSM. Using two field output variables of the enriched elements' nodes, PHILSM and PSILSM, LSM is used in XFEM to identify the crack and calculate the crack surface. The signed distance functions PHILSM (Φ) and PSILSM (Ψ) are used to characterize the crack surface and the first crack front [286]. For propagating cracks, the crack tip enrichment term is not accessible in XFEM. As a result, the propagating crack front will not stop inside the components. In a two-dimensional model, the crack segment for each failed element is a straight line, and in a three-dimensional model, it is a plane. We can determine the points whose PHILSM values are equal to 0 by using the PHILSM values of the four nodes. The coordinates of these points can be determined using the node coordinates. Furthermore, the increment crack length can be determined, and the total crack length can be computed by adding all these crack segments together. After explaining how to compute the fracture length in a two-dimensional model, the following section will explain how to calculate the crack area using PHILSM data. The elements in our cylindrical model are tetrahedral, with four nodes per element. It is explained in Figure 4.14 how to determine the coordinates of the intersecting location.

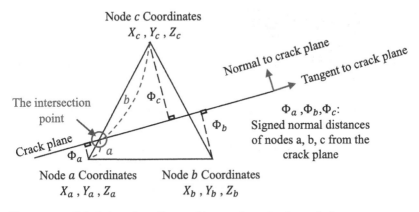

Figure 4.14 Calculation of coordinates of intersection points in crack plane

The PHILSM values of nodes a, b and c are Φ_a, Φ_b and Φ_c. Therefore, $a/b = |\Phi_a/\Phi_c|$. a and b are the distances between the points of intersection and node a and node c, respectively. As a result, the intersection point's coordinates can be written as

$$
\begin{aligned}
X &= X_a + (X_c - X_a) \cdot \left| \frac{\Phi_a}{\Phi_a - \Phi_c} \right|; \\
Y &= Y_a + (Y_c - Y_a) \cdot \left| \frac{\Phi_a}{\Phi_a - \Phi_c} \right|; \\
Z &= Z_a + (Z_c - Z_a) \cdot \left| \frac{\Phi_a}{\Phi_a - \Phi_c} \right|.
\end{aligned}
\tag{4.33}
$$

It should be emphasized that the intersection point exists only when the signs of the two points' PHILSM values are opposite. When two PHILSM values are both positive and negative, it means the two points are on the same side of the equation. The highlighted area in Figure 4.15 depicts a triangular fracture section. The triangle's sides a, b, and c can be computed using the coordinates of the points (point 1, point 2, and point 3)

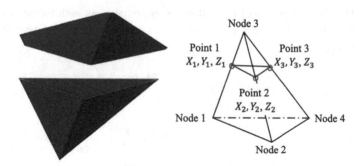

Figure 4.15 Projected area of a cracked triangular element in Abaqus

$$
\begin{aligned}
a &= \sqrt{(X_1 - X_2)^2 + (Y_1 - Y_2)^2 + (Z_1 - Z_2)^2}; \\
b &= \sqrt{(X_2 - X_3)^2 + (Y_2 - Y_3)^2 + (Z_2 - Z_3)^2}; \\
c &= \sqrt{(X_1 - X_3)^2 + (Y_1 - Y_3)^2 + (Z_1 - Z_3)^2}; \\
s &= \frac{a+b+c}{2}.
\end{aligned}
\tag{4.34}
$$

Heron's formula [302] can be used to calculate the crack area

$$
A = \sqrt{s(s - a)(s - b)(s - c)}
\tag{4.35}
$$

The triangle area is calculated using the Python function *myareaTri*, with the input values being the coordinate values of the triangle's three vertices. It is demonstrated how to compute the area of a quadrilateral fracture section. The sides and diagonals of the quadrilateral are a, b, c, d, e, and f, as illustrated in Figure 4.16.

$$a = \sqrt{(X_1 - X_2)^2 + (Y_1 - Y_2)^2 + (Z_1 - Z_2)^2};$$
$$b = \sqrt{(X_2 - X_3)^2 + (Y_2 - Y_3)^2 + (Z_2 - Z_3)^2};$$
$$c = \sqrt{(X_3 - X_4)^2 + (Y_3 - Y_4)^2 + (Z_3 - Z_4)^2};$$
$$d = \sqrt{(X_1 - X_4)^2 + (Y_1 - Y_4)^2 + (Z_1 - Z_4)^2};$$
$$e = \sqrt{(X_1 - X_3)^2 + (Y_1 - Y_3)^2 + (Z_1 - Z_3)^2};$$
$$f = \sqrt{(X_2 - X_4)^2 + (Y_2 - Y_4)^2 + (Z_2 - Z_4)^2}; \tag{4.36}$$
$$s_1 = \frac{a+b+e}{2};$$
$$s_2 = \frac{a+d+f}{2};$$
$$s_3 = \frac{c+d+e}{2};$$
$$s_4 = \frac{b+c+f}{2}.$$

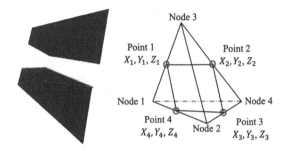

Figure 4.16 Projected area of a cracked quadrilateral element in Abaqus

The following are the areas of four triangles that have been formed

$$A_1 = \sqrt{s_1 \cdot (s_1 - a) \cdot (s_1 - b) \cdot (s_1 - e)};$$
$$A_2 = \sqrt{s_2 \cdot (s_2 - a) \cdot (s_2 - d) \cdot (s_2 - f)};$$
$$A_3 = \sqrt{s_3 \cdot (s_3 - c) \cdot (s_3 - d) \cdot (s_3 - e)}; \tag{4.37}$$
$$A_4 = \sqrt{s_4 \cdot (s_4 - b) \cdot (s_4 - c) \cdot (s_4 - f)};$$
$$A = \frac{A_1 + A_2 + A_3 + A_4}{2}$$

It is possible to determine the crack area A. The quadrilateral area is calculated using the function *myareaQuad*, with the input values being the coordinates of the quadrilateral's four vertices.

4.3.3 Damage Initiation Criteria

XFEM can be used to model the propagation of fatigue cracks along any route. The criteria are a VCCT-based fracture-based surface behavior. Because VCCT is based on the LEFM concept, it necessitates an initial crack in the model. A line portion in a 2D model or a shell part in a 3D model can be used to define the fracture. The part should be constructed in the Assembly module as an instance, with the direct cyclic step being the only step. If there is not a pre-defined crack, one must first be created in a static step [285]. The chosen damage start criteria govern fracture nucleation. After the fracture has been nucleated in the static stage, it can propagate in the direct cyclic step that follows. Abaqus offers six damage initiation criteria, all of which are based on the traction separation laws and are accessible for XFEM enhanced components. The damage process starts when the stress or strain levels satisfy the defined damage beginning criteria. Maximum nominal stress criterion (Maxs), Quadratic nominal stress criterion (Quads), Maximum principal stress criterion (Maxps), Maximum nominal strain criterion (Maxe), Quadratic nominal strain criterion (Quade), and Maximum principal strain criterion (Maxpe) [303], Maxe and Quade are assessed using three strain values: nominal strain in the pure normal mode (ε_n), first shear direction (ε_s), and second shear direction (ε_{ts}). These strain components' peak values are ε_n^{max}, ε_s^{max} and ε_{ts}^{max}.

$$f = MAX\left\{ \frac{\langle \varepsilon_n \rangle}{\varepsilon_n^{max}}, \frac{\varepsilon_s}{\varepsilon_s^{max}}, \frac{\varepsilon_{ts}}{\varepsilon_{ts}^{max}} \right\} \tag{4.38}$$

$$f = \left(\frac{\langle \varepsilon_n \rangle}{\varepsilon_n^{max}} \right)^2 + \left(\frac{\varepsilon_s}{\varepsilon_s^{max}} \right)^2 + \left(\frac{\varepsilon_{ts}}{\varepsilon_{ts}^{max}} \right)^2 \tag{4.39}$$

For ε_n, if $\varepsilon_n < 0$, $\varepsilon_n = 0$ and if $\varepsilon_n > 0$, $\varepsilon_n = \varepsilon_n$. To put it another way, the damage will not be caused by a pure compression load. Stress data, which are nominal stress in the pure normal mode (σ_n), in the first shear direction (σ_s), and in the second shear direction (σ_{ts}) [303], are used to assess Maxs and Quads. The tolerance f_{tol} should be set after the damage initiation criterion has been specified. The initiation only occurs after $1.0 \leq f \leq f_{tol} + 1.0$ is reached. While increasing the tolerance will speed up convergence, a large tolerance would surely impair accuracy and can result in many cracks. The Maxps and Maxpe criteria can only be applied to the XFEM enriched region's damage initiation. These two requirements determine a fracture plane that is perpendicular to the direction of the highest main stress or strain and cannot be user-defined. The Quade, Maxe,

Quads, Maxs damage initiation criteria can be utilized for cohesive components in the delamination process for composite materials as well as XFEM enriched elements. These four criteria can be used to forecast which direction the crack will expand in when predicting XFEM damage initiation: orthogonal to the local 1-direction or orthogonal to the local 2-direction. Because the stress and strain values in three directions are taken into consideration, the Quade and Quads criteria are more accurate than the Maxe and Maxs criteria. All six criteria can be used to provide tolerance to regulate accuracy at a point to compute stress or strain values, such as the centroid, crack tip, or combination, and temperature-dependent data. The Quade criterion is used as the damage start criterion in this work, and the f_{tol} is 0.05.

Damage assessment strategy
The initiation and development of fatigue cracks in Abaqus are dictated by the ΔG value at the crack tip, which is computed using VCCT. For instance, in a two-dimensional model, the crack length for stable cycle N is a_N. The crack will develop from a_N to $a_{N+\Delta N}$ when the energy release rate threshold G_{thresh} is reached. The crack will not develop if $G_{max} < G_{thresh}$. If $G_{max} > G_{pl}$ the number of cycles incremented (ΔN) is 1. If $G_{thresh} < G_{max} < G_{pl}$ the following procedure determines ΔN. The fatigue crack initiation criterion f is defined as [288]

$$f = \frac{N_0}{c_1 \Delta G^{c_2}} \geq 1.0 \qquad (4.40)$$

c_1 and c_2 are material constants and $\Delta G = G_{max} - G_{min}$. The fatigue crack development criterion is applied to all enriched elements that will fail. It specifies the minimum number of cycles that must be completed. When $G_{thresh} < G_{max} < G_{pl}$ da/dN is determined by Paris law

$$\frac{da}{dN} = c_3 \Delta G^{c_4} \qquad (4.41)$$

c_3 and c_4 are material constants. The update cycle ΔN is

$$\Delta N = \Delta a / \left(\frac{da}{dN} \right) \qquad (4.42)$$

It is worth noting that ΔN comes from the element with the fewest increment cycles and that this element will be broken at the conclusion of this stable cycle.

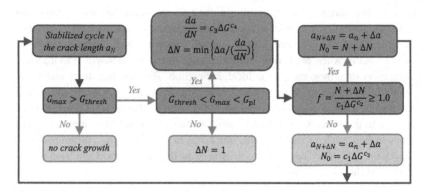

Figure 4.17 Flow chart of fatigue crack propagation and fatigue damage assessment based on the critical energy release rate

The increment crack length is given by Δa. The fatigue crack propagation rate for various directions is calculated using Abaqus, and Δa is chosen by the route with the highest fatigue crack propagation rate. If $N + \Delta N > c_1 \Delta G^{c_2}$ the onset criterion is not functioning, and the increment cycle number is defined by Paris law, which means the onset criterion is not working. Then

$$a_{N+\Delta N} = a_N + \Delta a \tag{4.43}$$

$$N = N + \Delta N \tag{4.44}$$

If $N + \Delta N < c_1 \Delta G^{c_2}$ it means that the current number of cycles derived from the Paris law does not match the fatigue onset criterion's minimum cycle requirement. As a result, given fracture length $a_{N+\Delta N}$ the stabilized number of cycles is equal to $c_1 \Delta G^{c_2}$. Then

$$a_{N+\Delta N} = a_N + \Delta a \tag{4.45}$$

$$N = N_0 = c_1 \Delta G^{c_2} \tag{4.46}$$

Repeating the above steps until the maximum number of cycles has been achieved or the crack has reached the enriched element's limits. The flow of data is shown in Figure 4.17.

4.3.4 Model Establishment and Simulation

All simulations were performed on a supposedly SLM-built cylindrical fatigue specimen. The fatigue specimen geometry is depicted in Figure 3.31. Due to the constraints of the μ-CT equipment, we will not be able to scan the entire specimen; instead, we will focus on the gage area, which is our region of interest (ROI). X-ray microcomputed tomography (μ-CT) scans produce pure 2D scans that are taken in angular increments. The increment has been set to $0.22°$ for every scan in this study. The scans are then uploaded into VGStudio Max 2.2, which stitches and joins the 2D scans using picture intensities and geometrical features to create a 3D shell representation. Because μ-CT has a high resolution, it can catch even the tiniest surface changes and interior porosity. If the original 3D shell representation generated by VGStudio Max 2.2 is used, the number of vertices will exceed 2.6 million, with several faces reaching 5.2 million, making the study impracticable. The 3D representation is saved as an STL (Standard Tessellation Language) file (see Figure 4.18) with VGStudio Max 2.2. Vertices and faces are used in an STL file to represent the geometry in question. Vertices are a collection of points found at geometrical features or discontinuities. The edges are then used to combine the vertices to make a face, which is subsequently linked together to produce a 3D representation. Tetrahedral elements are used in STL files, which means a face is connected to three vertices. To reduce the number of faces and vertices, numerous reductions and simplification techniques must be carried out. VGStudio Max 2.2 was used to perform the initial reduction procedure, which involved point reduction. By establishing a tolerance range, the point reduction process minimizes the number of points generated initially. If a given point A is inside the tolerance range of nearby points, it is eliminated, but the rest of the points are saved to retain the structure topography. MeshLab is used for the second simplification procedure, which employs the quadric edge collapse decimation technique [304]. This approach combines adjacent tetrahedral elements to generate a single tetrahedral element. This method reduces thumbprints of faces and vertices, allowing the simulation to continue. Nonetheless, the model's simplification techniques made it more prone to errors and deviations in findings when compared to a non-simplified 3D geometry. To solve this problem, mesh sensitivity analysis will be conducted in section (4.4.2) to determine the output deviation as a function of the mesh size. STL files are, in theory, shell representations of the geometry at hand, as shown in Figure 4.18.

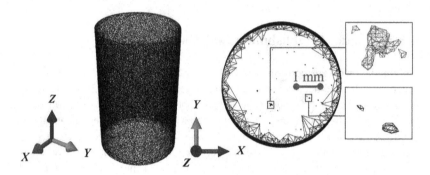

Figure 4.18   Tomographic surfaces of internal defects by STL representation showing various defect morphologies in a fatigue specimen based on μ-CT scans

An STL shell representation must be converted to a solid, and internal porosity must be reduced from the solid representation to imitate the consequences of internal porosity. Separating the cylindrical shell from interior porosities was accomplished using FreeCAD. Based on this, a boolean differential operator was applied to the solid cylindrical and solid differential porosities, resulting in a single solid cylindrical porous geometry that closely resembles those created during the SLM process. Finally, the solid cylindrical porous geometry will be saved as a STEP file, which can be imported as a component into Abaqus. The elastoplastic response of the as-built SLM sample will be simulated using the imported STEP file. The alloys AlSi10Mg and Ti-6Al-4V are applied, with elastic modulus E of 73.13 and 117.95 GPa as well as Poisson's ratio of 0.3. The damage initiation criterion is chosen to be the Quade criterion.

Simulated cyclic deformation test

In Abaqus, the direct approach will be utilized to mimic a stable cyclic response. The simulation will be set at 20 Hz for AlSi10Mg and 5 Hz for Ti-6Al-4V, and the test will be under stress control at a ratio of $R = -1$. Abaqus simulates a cyclic load using a periodic formulation based on the Fourier series, which is written as [305]

$$f(t) = A_0 + \sum_{n=1}^{\infty} [A_n \cos(n\omega(t - t_0)) + B_n \sin(n\omega(t - t_0))] \qquad (4.47)$$

where A_0 is the initial amplitude at a time t_0, A_n and B_n are time-dependent amplitudes, and the angular frequency ω is

$$\omega = 2\pi f \qquad (4.48)$$

To speed up the simulation process for continuum damage mechanics, the forward damage extrapolation method is used [306]. The relative energy release rate at the fracture tip characterizes the damage extrapolation approach for XFEM. The field output variables PHILSM, PSILSM, STATUSXFEM, and ENRRTXFEM are requested [307]. PHILSM and PSILSM are used to locate the crack position and front in the output database (ODB) file [282], and they are required to reveal the crack in the ODB file. The state of the enhanced elements is represented by STA-TUSXFEM. Damage has not begun if the element's STATUSXFEM value is 0, and if the value is 1, the element is completely failed [292]. The energy release rate values of the elements can be accessed using ENRRTXFEM. The discontinuous analysis is considered with $I_0 = 8$ and $I_R = 10$. I_0 is the frequency of checking the iteration residuals, and I_R is the beginning point of the convergence checks. The discontinuous condition plays a role in crack propagation, and raising the discontinuous parameters can increase efficiency and aid convergence [308]. I_A is the maximum number of cutbacks permitted in an increment, and it is set to 50. Because the crack initiation in Step-1 is so complex, increasing I_A is a must to avoid the error induced by the restricted number of cutbacks. Step-2 will be updated with the new general solution controls. Abaqus GUI can now incorporate the hard pressure-overclosure of the typical behavior and the VCCT fracture criterion as contact property choices for IntProp-1. The fatigue fracture criteria, on the other hand, can only be introduced by changing the keywords [305]. Each stage in Abaqus should have at least one accessible fracture criterion. Because phase 1 is a static step with no periodic stress, the fatigue fracture criteria do not apply. As a result, the VCCT fracture criterion is established as the Step-1 fracture criterion. Because the maximal energy release rate in Step-1 is significantly lower than G_c the fracture in Step-1 will merely begin rather than expand [305]. In the Interaction module, an XFEM crack named Crack-1 is produced. The center area is the crack domain, the crack location is unknown, and the contact property is created IntProp-1. It is permissible for cracks to grow. The VCCT or improved VCCT fracture criterion can only be created in Abaqus' GUI. Editing the keywords should be used to build the fatigue fracture criterion. Because there is a small pore inside the component, only the free method is used in this work. C3D4 is the element type (4-node linear tetrahedron) and tet is the shape of the element. The overall dimension is 0.5 mm. The crack opens, and the number of failed elements grows as the cycle number increases, as illustrated in Table 4.2.

Table 4.2 Evolution of fatigue crack and failed elements with the number of cycles of Ti-6Al-4V at 600 MPa

Cycle No.	The crack opening	The failed elements
$N=$ 14599		
$N=$ 17872		
$N=$ 22142		
$N=$ 58590		
$N=$ 81039		

Because the pore is the specimen's weakest point, damage begins there. At cycle 81039, the fracture spreads throughout the specimen, causing the structure to break. The fracture area and cycle number for each selected frame are extracted using the Python script after the simulation. The fracture propagates slowly at first in Figure 4.19, but da/dN progressively increases, and the growth speed accelerates. Because the energy release rate has exceeded the top limit G_{pl} towards the end of the curve, the slope is exceptionally steep, and the curve begins to follow the fatigue fracture development curve in Region III, rather than the Paris law in Region II.

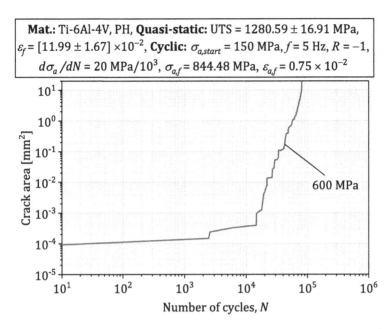

Mat.: Ti-6Al-4V, PH, **Quasi-static:** UTS = 1280.59 ± 16.91 MPa, ε_f = [11.99 ± 1.67] ×10^{-2}, **Cyclic:** $\sigma_{a,start}$ = 150 MPa, f = 5 Hz, R = −1, $d\sigma_a/dN$ = 20 MPa/10^3, $\sigma_{a,f}$ = 844.48 MPa, $\varepsilon_{a,f}$ = 0.75 × 10^{-2}

Figure 4.19 Evolution of crack area in the orthogonal direction to load application for a specimen of Ti-6Al-4V at 600 MPa stress amplitude

In Figure 4.20, the fatigue crack area evolution is presented as a function of the number of cycles at different stress amplitudes for AlSi10Mg. In all shown stress amplitudes, the fatigue crack growth area is proceeding at a very slow rate before 1E3 cycles. For 140 MPa, an abrupt increase in the crack area takes place at around 3E3 cycles, followed by a stable horizon until around 4E4 cycles. After this, the unstable crack propagation zone proceeds towards ultimate failure giving the final number of cycles to failure. For 130 and 120 MPa, the first significant rise in fracture area happens between 5E3 and 6E3 cycles. For 110 and 100 MPa, this happens between 8E3 and 2E4. Although the simulations were conducted uniformly in decreasing stress amplitudes of 10 MPa, the response of the simulated crack area is highly nonlinearly proportional to the increase in the number of cycles to failure. For 110 and 100 MPa of stress amplitude, the unstable fracture does happen until after 1E5 cycles.

The result can be used to plot crack propagation rate curves and study the influence of loading level and testing frequency on fatigue crack propagation rate. The

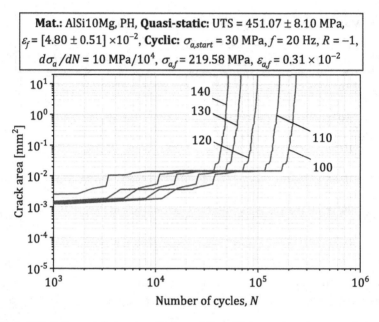

Figure 4.20 Influence of stress amplitude on fatigue lifetime and crack area evolution in AlSi10Mg

author supposes that the resulting Paris exponent would be similar; however, the Paris coefficient is expected to be load-dependent and will shift towards higher thresholds for lower stress amplitudes. The dependency of fatigue crack area evolution in Ti-6Al-4V is shown in Figure 4.21. The crack area evolution rate is slow and steady until just after 1E3 cycles, where the first acceleration fatigue crack area evolution starts to accumulate. The accelerated crack area evolution, which is unstable, begins at 5E3 cycles or lower for stress amplitudes 800, 750, and 700 MPa. For 650 and 600 MPa of stress amplitude, this phase starts between 5E3 and 1E4 cycles. It is obvious that although the simulations take place at uniformly distributed stress levels, the evolution of the crack area is not linearly proportional. Based on these calculations, crack propagation rate curves can be used to conclude about Paris constants of this studied specimen geometry in dependence on the applied load. Figure 4.22 represents a comparison between simulated fatigue lifetime and the experimentally obtained values.

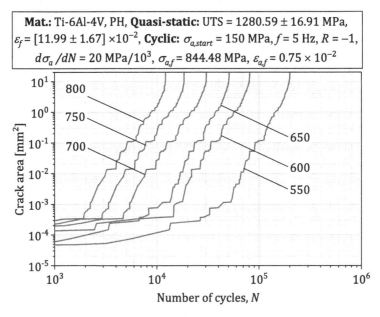

Mat.: Ti-6Al-4V, PH, **Quasi-static:** UTS = 1280.59 ± 16.91 MPa, ε_f = [11.99 ± 1.67] ×10^{-2}, **Cyclic:** $\sigma_{a,start}$ = 150 MPa, f = 5 Hz, R = –1, $d\sigma_a/dN$ = 20 MPa/10^3, $\sigma_{a,f}$ = 844.48 MPa, $\varepsilon_{a,f}$ = 0.75 × 10^{-2}

Figure 4.21 Influence of stress amplitude on fatigue lifetime and crack area evolution in Ti-6Al-4V

For both AlSi10Mg and Ti-6Al-4V, a good agreement can be found between experiment and simulated values. Hence, the applied methodology does not only visualize the failure process from subsurface defects but also represents an accurate depiction of the nonlinear relationship between stress amplitude and the number of cycles to failure for two distinctively different alloying systems, which have unique fatigue failure mechanisms.

4.4 Modeling of Cyclic Deformation

4.4.1 Continuum Material Models

A continuum material model is a mathematical depiction of a material's response to external loads, especially in the case of elastoplastic response under cyclic or quasi-static loading. A yield criterion and a hardening model are part of each

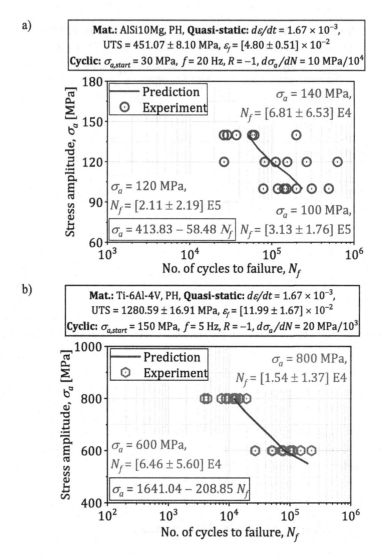

Figure 4.22 Projections of experimental fatigue data and the calculated lifetimes using the extended finite element method method: a) AlSi10Mg; b) Ti-6Al-4V

continuum material model [176, 309]. Crystal slip is the source of plasticity in crystalline materials. Metals are often polycrystalline, meaning they are made up of many crystals with atoms stacked in a regular pattern [310]. The yield condition, flow rule, and hardening law are all part of the continuum constitutive plastic model's description [176, 311]. We can envision plastic deformation taking place as shown in Figure 4.23a and b if we depict the crystallographic structure of a tiny portion of a single grain by planes of atoms, as shown in Figure 4.23a and b. This is a crystallographic slip. Unlike elastic deformation, which simply involves stretching interatomic connections, slip necessitates the breaking and re-forming of interatomic connections as well as the relative displacement of one plane of atoms to another [309].

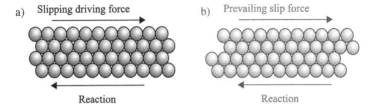

Figure 4.23 Slipping of a single grain crystallographic structure: a) before deformation; b) after deformation [310]

Except at the crystal's boundaries, the structure remains intact after shearing it from configuration Figure 4.23a to configuration Figure 4.23b. Only two of the most important phenomena in macroscopic plasticity can be seen in Figure 4.23: 1/ Plastic slip does not lead to volume change; this is the plasticity incompressibility condition; 2/ Plastic slip is a shearing process; hydrostatic stress is frequently thought not to affect slide at the macrolevel. 3/ Plastic yielding in polycrystals is frequently an isotropic phenomenon [176, 310]. Aside from the microstructure, a simulation program relies on accurate and detailed material parameters to generate results. The kinematic hardening parameter and the isotropic hardening parameter are critical material parameters for fatigue behavior modeling. Because the flow curve alone cannot adequately explain the material's properties, both hardening parameters are used to provide a detailed description of the material. There are several methods for obtaining such material parameters.

Yield criterion

The yield condition of the material helps in determining whether the material is undergoing plastic strain deformation or reversible elastic strain. The back stress and other factors determine the general three-dimensional yield condition under cyclic loading [176, 181]. As a result, the yield surface F, of a material can be expressed as a function of

$$F = f(\sigma, x, p) \tag{4.49}$$

where p represents the plastic strain, and x is the back stress [181]. The yield condition could be interpreted into two states [176]

$$F(\sigma, x, p) \leq 0 \tag{4.50}$$

→ elastic deformation

$$F(\sigma, x, p) > 0 \tag{4.51}$$

→ plastic deformation

The yield condition for a general cyclic loading test is as follows [176]

$$F = \left(\frac{3}{2}(\sigma' - x') : (\sigma' - x')\right)^{1/2} - \sigma_y(p) \tag{4.52}$$

$$\sigma_y(p) = \sigma_{y0} + \sigma(p) \tag{4.53}$$

where σ' is a deviatoric tensor of the stress component, x' is the deviatoric tensor of the back stress [181], $\sigma_y(p)$ is the yield stress as a function of accumulated plastic strain, σ_{y0} is the initial yield stress, and $\sigma(p)$ is the increment in the yield stress as a function of accumulated plastic strain [176]. In the case of 3D models, the bold notations of stress, back stress [181], and cumulative strain reflect a tensor notation is used, which is a universal form [176, 304, 310, 312].

Flow rule

The flow rule defines the development of the yield surface in the stress space, in the case of isotropic hardening, and translation, in the case of kinematic hardening, depending on the load history, which will be divided into microscopic steps [176,

181]. As a result, the incremental increase in the plastic strain because of each step
is

$$d\boldsymbol{\varepsilon}^{pl} = d\lambda r(\boldsymbol{\sigma}, \boldsymbol{x}, \boldsymbol{p}) \tag{4.54}$$

where $d\boldsymbol{\varepsilon}^{pl}$ is the constitutive plastic strain, $d\lambda$ is a constancy parameter, defining
the amount of the yield surface's development or translation, while $r(\boldsymbol{\sigma}, \boldsymbol{x}, \boldsymbol{p})$ is a
function that maps the course of evolution or translation [176]. According to the
Drucker-Prager criterion [176, 181, 313], the flow can be stated as [176]

$$d\boldsymbol{\varepsilon}^{pl} = d\lambda \frac{\partial F(\boldsymbol{\sigma}, \boldsymbol{x}, \boldsymbol{p})}{\partial \boldsymbol{\sigma}} \tag{4.55}$$

Under the premise of von Mises yield [176, 177, 181]; Figure 4.24, the yield surface
expands in a normal way evenly in all directions [176, 177]. This flow rule is referred
to as the normal rule [309, 310].

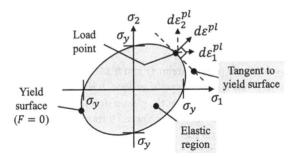

Figure 4.24   Two-dimensional projection of the yield surface and the normal to the tangent
to the yield surface

It is worth noting that flow rules are divided into two groups [176]: 1/ associative
flow and 2/ non-associative flow

$$d\varepsilon^{pl} = d\lambda \frac{\partial F(\boldsymbol{\sigma}, \boldsymbol{x}, \boldsymbol{p})}{\partial \boldsymbol{\sigma}} \tag{4.56}$$

$\rightarrow$ associative flow rule

$$d\varepsilon^{pl} = d\lambda \frac{\partial Q(\sigma, x, p)}{\partial \sigma} \tag{4.57}$$

→ non-associative flow rule

Hardening law

To complete the fatigue simulation, we need the kinematic hardening parameter and isotropic hardening parameter sets. All materials harden under stress. Plastic deformation occurs when the stress is greater than the yield point. The multiaxial yielding is the same in a perfectly plastic cylinder. Yield surface is defined as

$$F\left(\sigma_{ij}, H_i\right) = 0 \tag{4.58}$$

H is the hardening modulus. At first yield, H is 0. Isotropic hardening can be assumed to be a function of the equivalent plastic strain p. The yield surface is changed in size but not in shape and position. Chaboche suggested the following form [314]

$$F = \sigma_{y0} + Q\left(1 - e^{-bp}\right) \tag{4.59}$$

σ_{y0} is yield stress at zero plastic strain, Q and b are material parameters. Q is the maximum change in the size of the yield surface. b is defined as the rate at which the size of the yield surface changes as the plastic strain develops. In the simulation, both parameters are necessary and are acquired through genetic algorithm experiments [315]. This is an integral part of the Ottosen-Stenstroem-Ristinmaa approach equipped with the damage rule of Lemaitre and Chaboche [316]. When the load is increased, the yield surface expands [176, 177, 181]. The isotropic hardening cannot exhibit the Bauschinger effect, which is that hardening contributes to softening effect in the opposite direction. The yield surface does not vary in size or form, but it translates in stress space due to the Bauschinger effect. The components of Chaboche's kinematic hardening are represented by [314]

$$X = \sum_{i=1}^{m} X_i \tag{4.60}$$

m is the number of back stress components used in order to obtain an acceptable material characteristic under cyclic loading. In metals, there is a combined hardening equation that incorporates both kinematic and isotropic hardening [176, 181, 310]. Ziegler's rule and Prager's rule are two different instances of linear back stress development [181]. Another way to explain this is through the Armstrong-Frederick (AF) or Chaboche laws. Typically, hardening rules can be classified into three types: linear, ideal, and arbitrary (nonlinear). Nonlinear kinematic and nonlinear isotropic hardening formulas are employed to ensure cyclic uniaxial material behavior predictions; see Figure 4.25 [176, 181]. Armstrong and Frederick's nonlinear kinematic and isotropic hardening are listed respectively as follows [176, 181]

$$dx = \frac{2}{3} C d\varepsilon^{pl} - \gamma x dp \tag{4.61}$$

$$\sigma_y(p) = \sigma_{y0} + Q\left(1 - e^{-bdp}\right) \tag{4.62}$$

where dp is the accumulated plastic strain, which is acclaimed as

$$dp = \left|d\varepsilon^{pl}\right| = \left[\frac{2}{3}\varepsilon^{pl} \cdot \varepsilon^{pl}\right]^{1/2} \tag{4.63}$$

Where C, γ, Q and b are material parameters that need to be established based on the outcomes of experiments. The value of the absolute plastic strain increment is simplified, as shown by Awd et al. [176, 304]. Variable amplitude tests are used to identify the kinematic hardening parameter [317]. Symmetrical stress cycles allow solely for cyclic stability. This connection is given as stress amplitude vs. strain amplitude

$$\frac{\Delta\sigma}{2} = X_{max} + k = \frac{C}{\gamma} tanh\left(\gamma \frac{\Delta\varepsilon_p}{2}\right) + k \tag{4.64}$$

The progress of ratcheting occurs theoretically as soon as the mean stress is not zero and the stress range is more than $2k$, where k is the initial size of the yield surface. The progressive strain of every cycle can be expressed by

$$\Delta \varepsilon_p = \frac{1}{\gamma} \left[\frac{\left(\frac{C}{\gamma}\right)^2 - (\sigma_{min} - k)^2}{\left(\frac{C}{\gamma}\right)^2 - (\sigma_{max} - k)^2} \right] \qquad (4.65)$$

The values of k, C are determined using variable amplitude tests. The elastic domain determines the value of k. Because of the initial kinematic hardening modulus C, the yield stress, in this case, is slightly larger than σ_y.

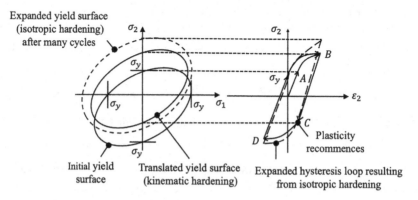

Figure 4.25 Illustration of yield surface expansion and translation in the three-dimensional parameter space

The curve fitting method is used to find the value of C and γ in the relationship. The values of C and γ will be entered into the Abaqus software to give the material a kinematic hardening characteristic. Other procedures, such as genetic algorithm [318] or two-stage loading [319], can be used to acquire kinematic hardening parameters in addition to multi-amplitude strain-controlled cyclic loading. In various materials, pure kinematic or pure isotropic hardening is not to be found. Changes in location and expansion of the yield surface are common. When location is constant, that indicates pure isotropic hardening, and when the expansion of the yield surface does not take place, it means pure kinematic hardening [320]. The progression of isotropic hardening is represented here by the development of R as a function of p. The increasing trend indicates cyclic hardening, whereas decreasing trend indicates cyclic softening [317]. Because R evolves, the equation resembles the isotropic hardening relationship

$$dR = b(Q - R)dp \qquad (4.66)$$

which corresponds to a regime of stabilized cycles, and b indicates the speed of the stabilization. Each uniaxial cycle's relation and application are integrated as follows

$$\sigma_{max} = X_{max} + k + Q[1 - exp(-bp)] \qquad (4.67)$$

If the plastic strain range $\Delta\varepsilon^{pl}$ is considered constant and X_{max} is also assumed to be constant, the Nth cycle in a stabilized case is

$$\frac{\sigma_{max} - \sigma_{max}^0}{\sigma_{max}^s - \sigma_{max}^0} = 1 - exp\left(-2b\Delta\varepsilon^{pl}N^{th}\right) \qquad (4.68)$$

The maximum stress at the first loop and the stabilized loop, respectively, is σ_{max}^0 and σ_{max}^s in a strain-controlled analysis. The maximum stress of a random loop is σ_{max}. The value of b is calculated using the curve-fitting approach once the data has been collected. The value of Q is obtained by

$$\sigma^0 = \sigma_0 + Q\left(1 - e^{-b\varepsilon^{pl}}\right) \qquad (4.69)$$

σ^0 is the yield stress for the Nth cycle, σ_0 is the yield stress at zero plastic strain, Q and b are fitted in this Equation 4.69 [56].

4.4.2 Implementation and Solution Optimization

The approach adopted here to quantify internal stresses rely on four main parts: 1/ Microcomputed tomography: Acquiring the real geometry as-built by X-ray microcomputed tomography μ-CT and extracting it into FE meshes with their as-built morphology as realistic as possible; 2/ Material behavior and cyclic loading experiments: Using a multiple back stress cyclic plastic models based on Armstrong-Frederick kinematic hardening rule under consideration of isotropic expansion of the yield surface and performing calibration experiments of isotropic and kinematic hardening parameters; 3/ Finite element method: In the FEM application, efficient element size is determined through mesh sensitivity analysis and application of realistic boundary conditions which reflect the testing conditions;

4/ Fatigue damage relation: Based on the critical shear strain identified in the FEM analysis, Fatemi-Socie model is calibrated, and cycles to failure of the specimen are deduced. An overview of the flow of data and corresponding analysis steps are found in Figure 4.26. Specimens were scanned before testing at Nikon X TH 160 equipped with a 160 kV gun and an amorphous silicon-based digital detector with a 1024×1024-pixel count that corresponds to 127 μm pixel pitch. On VGStudio Max 2.2, the raw data of μ-CT is processed in order to extract the external surfaces of the specimens and internal surfaces of the pores. Python-based CAD software was used where pore surfaces in relation to the bulk specimen are stitched and joined. If a vertex is outside the tolerance range from the neighboring vertices, the vertex is deleted, while the rest of the points are preserved to save the structure topography. A second filter is conducted on MeshLab using a quadric edge collapse decimation algorithm where neighboring tetrahedral elements forming a single surface are joined to form a single element. The quasi-static test is applied to calibrate isotropic hardening parameters b, Q.

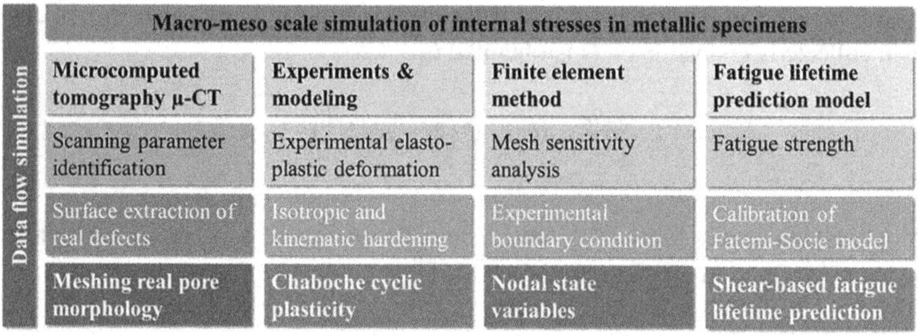

Macro-meso scale simulation of internal stresses in metallic specimens			
Microcomputed tomography μ-CT	Experiments & modeling	Finite element method	Fatigue lifetime prediction model
Scanning parameter identification	Experimental elasto-plastic deformation	Mesh sensitivity analysis	Fatigue strength
Surface extraction of real defects	Isotropic and kinematic hardening	Experimental boundary condition	Calibration of Fatemi-Socie model
Meshing real pore morphology	Chaboche cyclic plasticity	Nodal state variables	Shear-based fatigue lifetime prediction

Figure 4.26　Flow chart of the applied simulation scheme on the macro-mesoscale

The load increase tests calibrate the kinematic hardening parameters K, C, and γ with respect to the requested number of back stresses necessary to capture hysteresis translation in the three-dimensional space. The proposed technique allows the identification of isotropic and kinematic hardening everyone on its own without interaction since two independent procedures are used. An incremental increase of stress is applied on Abaqus, where stress amplitude levels are predefined and compared to the measured plastic strains from the load increase test so parameters of the isotropic hardening rule can be identified. The fatigue lifetime prediction model based on Fatemi-Socie damage depends on the maximum shear strain developed under simulated cyclic loading.

Isotropic hardening parameter calibration
A prior trial in Abaqus had suggested that the flow curve is not enough for simulating the system since the cyclic hardening effect and the Bauschinger effect are not captured. Except for applying the flow curve solely, the Abaqus material characteristic setup can provide the combined hardening parameter. This is collected during a numerical experiment. The numerical experiment is a tensile test, with load regulated by an increase in amplitude. With the support of these numerical experiments, only tensile and stress-controlled increasing load tests were included in the parameter optimization. Equation 4.68 defines the development of the isotropic hardening parameter [321]. The fatigue regime that is focused on here is the HCF. With a very low proportion of plastic strain response, fatigue tests must be stress-controlled. The isotropic hardening parameter can be derived through the simulation of a load increase numerical test. Uniaxial tensile stress-strain graphs and numerical experiments are used to calibrate the isotropic and kinematic hardening parameters [321]

$$\sigma = \sigma_0 + Q\left(1 - e^{-b\varepsilon^{pl}}\right) + \frac{C}{\gamma}\left(1 - e^{-\gamma\varepsilon^{pl}}\right) \tag{4.70}$$

The kinematic parameter is ignored while calibrating the isotropic hardening parameter. Additionally, while forecasting the kinematic hardening value, the isotropic hardening portion is ignored. Nonetheless, this technique is not mentioned in the literature. To decrease the number of experiments necessary previous work points to the use of simulation to calibrate the isotropic hardening parameter [322]. The trial-and-error approach is unable to calibrate the isotropic hardening value. Curve fitting is thus employed. Only the flow curve defines the material characteristic in this strain-controlled load rise simulation optimization. Series of data sets are generated by various amplitudes. Thus, the isotropic parameter can be calibrated via a regression technique. The hardening parameter is also double-checked through simulations and experiments. The basic goal is to compare the initial stress amplitude to the newly stabilized stress amplitude. First, a basic tensile test is run, and the material is calibrated using the flow-only curve, as shown in Figure 4.27.
    The flow curve is calibrated using the critical resolved shear stress, as the fatigue yield stress is significantly less than the macroscopic yield stress in the uniaxial tensile test. The boundary condition is used to reference points with a continuous distribution to the lower surface of the model. Stress is imparted on the top surface. The simulation results are shown in Figure 4.28, where SSY stands for small-scale yielding. The simulation produces an outcome with a small error such that further simulations in a cyclic setup are adequate. Uniaxial tensile testing

a)

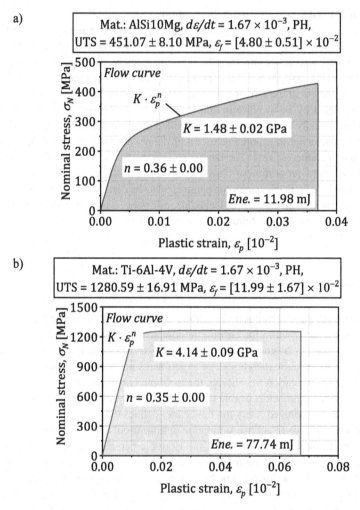

b)

Figure 4.27 Uniaxial tensile test showing the flow curve of selective laser melted: a) AlSi10Mg; b) Ti-6Al-4V

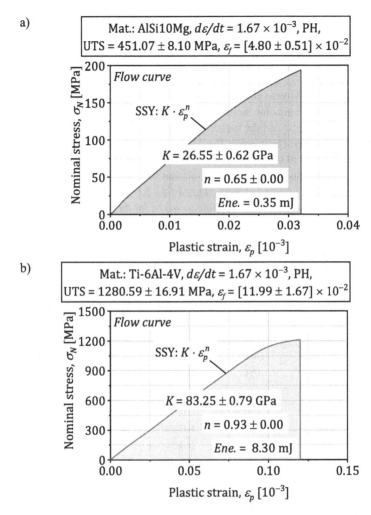

a)

Mat.: AlSi10Mg, $d\varepsilon/dt = 1.67 \times 10^{-3}$, PH,
UTS = 451.07 ± 8.10 MPa, $\varepsilon_f = [4.80 \pm 0.51] \times 10^{-2}$

Flow curve

SSY: $K \cdot \varepsilon_p^n$

$K = 26.55 \pm 0.62$ GPa

$n = 0.65 \pm 0.00$

Ene. = 0.35 mJ

b)

Mat.: Ti-6Al-4V, $d\varepsilon/dt = 1.67 \times 10^{-3}$, PH,
UTS = 1280.59 ± 16.91 MPa, $\varepsilon_f = [11.99 \pm 1.67] \times 10^{-2}$

Flow curve

SSY: $K \cdot \varepsilon_p^n$

$K = 83.25 \pm 0.79$ GPa

$n = 0.93 \pm 0.00$

Ene. = 8.30 mJ

Figure 4.28 Uniaxial tensile test simulation result with a small-scale yield: a) AlSi10Mg; b) Ti-6Al-4V

will also offer the strain range of the additional strain-regulated amplitude cyclic test. AlSi10Mg uniaxial tensile simulation reached ~ 180 MPa, whereas Ti-6Al-4V reached ~ 1200 MPa. The magnitudes tested are significantly above the intended fatigue experiment's load levels. From the simulation, the corresponding strain range can be calculated. For every material, a wide range of strain levels is applied with larger increments. Moreover, a restricted range is formed with smaller increments. For simulation, the genuine strain range for AlSi10Mg is 0.0012 to 0.0020 with an increase of 0.0002, and the higher increment range is 0.0012 to 0.0036 with an increment of 0.0006. The genuine Ti-6Al-4V strain range in simulation is 0.0060 to 0.0100 and 0.0040 to 0.0120 with increments of 0.0010 and 0.0020, respectively. Using 5 distinct values of amplitude levels can make the simulation simpler. Only one strain or stress amplitude was used in Abaqus for a single step at a time. Several steps of the simulation make it possible to adjust for the varying strain amplitude. For example, in each simulation, the 5-strain amplitude is simulated; each simulation includes five further steps. Each step contributes to one strain amplitude. Additionally, the duration of the step will be set as the number of complete cycles. Every phase has five complete cycles. This is shown in Table 4.3.

Table 4.3 Optimization of the finite element model based on true strain simulations to fit isotropic hardening parameters

True strain amplitude	AlSi10Mg (low)	AlSi10Mg (high)	Ti-6Al-4V (low)	Ti-6Al-4V (high)
Step time(s)	0.25	0.25	1	1
Step 1	0.0012	0.0012	0.0060	0.0040
Step 2	0.0014	0.0018	0.0070	0.0060
Step 3	0.0016	0.0024	0.0080	0.0080
Step 4	0.0018	0.0030	0.0090	0.0100
Step 5	0.0020	0.0036	0.0100	0.0120

For the simulation data fitting, isotropic hardening is the main objective function. Plastic hardening is specified to be isotropic exclusively. Every cycle has 40 iterations, so every step is separated by 0.00125 seconds for AlSi10Mg and 0.005 seconds for Ti-6Al-4V. Every step has 200 iterations, and every five-cycle affects five strain amplitudes. Simulating 1000 iterations will provide both a simple simulation and an acceptable outcome. The evolution of stress is shown. See Figure 4.29 below as a function of strain. When establishing an isotropic hardening parameter relationship, the stress amplitude must be evolutionary for accurate computation; however,

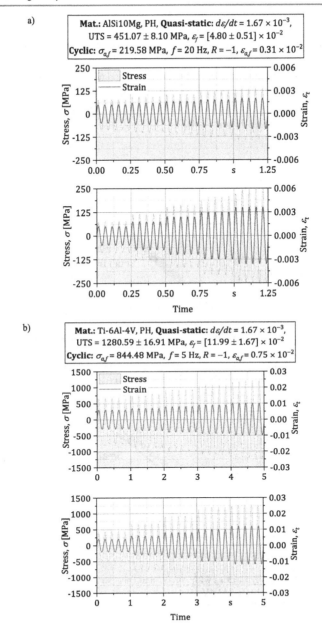

Figure 4.29 Controlling strain amplitude in a simulation to optimize the evolution of stress in the isotropic mechanism in: a) AlSi10Mg; b) Ti-6Al-4V

the laboratory experiment is stress controlled. Therefore, stress is a response rather than a control parameter in the numerical experiment. The simulation shows a clear difference in stress amplitude between the stabilized loop and the initial loop. This simulation's strain range is calculated using a uniaxial tensile test simulation result. Very near to the yield point, the nonlinearity is significantly increased. The greater increment offers a better assessment of the isotropic hardening parameter, which yields a larger range; see Figure 4.30.

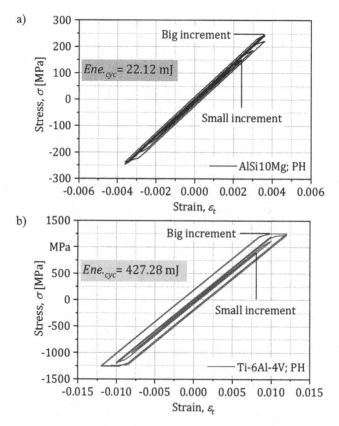

Figure 4.30 Isotropic hysteresis with small and big strain increments according to Table 4.3: a) AlSi10Mg; b) Ti-6Al-4V

Both simulations exhibit the isotropic hardening effect rather well, so the setup is feasible. The stabilizing procedure, however, happens in only one loop as it is in simulation. The y-side of Equation 4.68 has at least three separate maximum stress points. Half loop stabilization is the setting of the material characteristic. Cyclic loading can be stabilized after the first complete cycle. Strain amplitude and measurement inaccuracy will affect the ultimate result of the experiment. The stabilized loop is essential to find, and estimate based on these numerical experiments. To pick a certain iteration point for every single step, it is of primary importance to keep track of the maximum stress amplitude of the first loop and the calibrated loop. Maximum amplitude in a numerical loop is selected for the computation, which must be equivalent to the stress amplitude found in the stabilized loop in the experiment. This is only a relative factor necessary to facilitate the computation. Additionally, surpassing yield strain corresponding to the resolved shear stress necessitates several loops to be stabilized. The relative difference between the first three numerical loops is proportionally used. This is due to the extensive change in plastic that takes place during the initial cycles. Using the amplitude of the first, second, and third numerical loops, the y-side of the equation is calculated. b is calibrated by graphical analysis; therefore, a simple initial estimate can be performed [321]

$$\frac{\sigma_{max} - \sigma_{max}^0}{\sigma_{max}^s - \sigma_{max}^0} = 1 - \exp\left(-2b\Delta\varepsilon^{pl}N^{th}\right) \rightarrow y = 1 - \exp(bx) \qquad (4.71)$$

The value of b can be determined by numerical data fitting. Hence, Q is possible to get from [317] when the kinematic hardening part is omitted from the analysis in Equation 4.67. Thus, the parameters of isotropic hardening of Equation 4.71 should be as mentioned in Table 4.4.

Table 4.4 Isotropic hardening parameters

	AlSi10Mg	Ti-6Al-4V
b	52	8
Q	860	15100

Kinematic hardening parameter calibration

In Abaqus, C and γ are the simulation's essential values for kinematic hardening. The relationship is established with [323]

$$\frac{\Delta\sigma}{2} - k = \frac{C}{\gamma}tanh\left(\gamma\frac{\Delta\varepsilon_p}{2}\right) \qquad (4.72)$$

According to the literature [323], experiments should be conducted at a variety of preset stress amplitudes with a rising tendency. The stress amplitude, or corresponding plastic strain, is determined using just the half-life cycle to represent the stabilized loop. However, a different perspective is adopted in this work. The stress amplitude does not fluctuate between specified levels but increases gradually and steadily within a defined range according to the load increase test (LIT). The number of stress levels specified within a range of LIT is defined to be more than sufficient to establish a relation about the evolution of kinematic hardening. The purpose of this is to guarantee that the hardening substance is saturated over the investigated range of the LIT. Additionally, such an experiment provides a substantially larger number of data points for parameter calibration. As opposed to the experiment concepts described in the literature [323], which offer just a few data points, the LIT idea can provide continuous data from start to finish. It enables a more precise estimate and allows for an in-depth examination of the effects of the number of back stresses on the simulation outcome. According to the literature, increasing the number of pairings produces a more accurate result [323]. A redesigned experiment approach is offered. The hardness parameters change according to the combination used. Additionally, Abaqus enables material calibration with a maximum of ten material kinematic hardening parameter pairs. The simulation is proposed with a variable number of pairs of kinematic hardening parameters. A comparison of hardening pairs setups is performed to ascertain the optimal configuration for each material.

In the experimental data of the LIT, the behavior of AlSi10Mg is observed to have a progressive deformational strain from 30 MPa to roughly 100 MPa, from 100 to 150 MPa, and the strain evolution is far smaller compared to the earlier stages. Ti-6Al-4V experiences a gradual deformation evolution from around 700 to 800 MPa. The deformation rate increases a little, and the mechanism transition point is around 800 MPa. When exceeding such stress amplitude, the specimen elongates massively and eventually fails. This is extremely critical in choosing the optimal number of pairs of kinematic hardening parameters. Additionally, separating stress amplitude range for a certain set of kinematic hardening parameters. The kinematic hardening parameter relation is in Equation 4.72. The custom nonlinear fitting function which is applied here is

$$y = A \cdot tanh(Bx) + k \qquad (4.73)$$

The value of k is equal to k in Equation 4.72, B is equal to γ, and C can be calculated by A times B. The values of one hardening pair can be found in Table 4.5. Regression options will be determined with the minimum deviation with all single variables. The regression procedure, however, can only produce the most unbiased parameters. Target stress amplitude is one of the essential aspects for determining fatigue limits. This is a critical regression factor to accurately define the proper hardening parameters. To better fit the curve, the quantity B in Table 4.5 is calibrated by observing the mean strain values in the fatigue test. The material properties cannot be precisely specified by one pair of hardening parameters. The *tanh* functions give precise material behavior only in the target stress amplitude range. The *tanh* function cannot provide a certain outcome unless the fitting is restricted to the target stress amplitude range. It cannot represent a broad spectrum of parameter ranges of various stress amplitudes. Using the curve fitting results, it can be shown that the assumption of one hardening parameter cannot produce an appropriate simulation result. The separation of the fatigue damage mechanism transition region is based on that. The separation of regions for distinct pairs is the sole vital point that has the greatest influence on the parameter estimation process. Based on the previous observation, AlSi10Mg mechanism transition region into the LCF region is approximately 150 MPa, and Ti-6Al-4V is approximately 700 MPa. A regression was done below and above this stress amplitude zone.

Table 4.5 Hardening parameters' curve fitting function

	AlSi10Mg	Ti-6Al-4V
A	219.65 ± 0.71	879.22 ± 10.31
B	780 ± 0	1450 ± 0
k	-60.93 ± 0.51	-29.79 ± 6.56
Reduced Chi-sq	134.78	12803.88
R-**square**	0.91	0.69

In Ti-6Al-4V, the fitting function is able to correct for the ascending region, the transition, and the nearby fitted range. But the rising slope region has a big discrepancy in the variance to mean values. This location in the hardening parameter curve fitting causes it to be less accurate than expected. The curvature segmentation is better in the AlSi10Mg. Assumptions of three pairs produce a reliable material hardening parameter calibration for Ti-6Al-4V. However, the rising slope in the ascending zone has a degree of material calibration error. The curve fitting process delivers a satisfactory statistical and graphical result for AlSi10Mg and Ti-6Al-4V. The simulation could correctly generate four sets of kinematic hardening parameters for Ti-6Al-4V. The fitting of five pairs of kinematic hardening parameters was satisfactory for AlSi10Mg. A more complex material characteristic can be achieved by more pairs of parameters. Simulation efficiency, however, will suffer from overfitting. A comparison of resulting hysteresis loops will be compared up to 10 back stresses for AlSi10Mg and Ti-6Al-4V. The bias in the simulated and calibrated parameters stems from the bias of the experimental data for both alloys. The transition slope changed from test to test; so curve fitting was relocated to the transition region. Therefore, the calibration and simulation procedure was carried out in accordance with representative experimental values that are close as possible to the average experimental characteristics across several experimental trials for each type of test. A similar concept is applied in micromechanics [324], where a representative volume element is statistically constructed to mimic macroscopic material behavior. Aside from these unanswerable errors, curve fitting will always need to be applied to two distinct regions of the LIT, which are above and below the transition point for the HCF regime. Table 4.6 lists 10 pairs of kinematic hardening for AlSi10Mg and Ti-6Al-4V, respectively. All pairings of kinematic hardening parameters are simulated for the optimization of the simulation of both alloys.

Optimization of kinematic pairs
All the simulation results are compared to the experimental data for strain amplitudes and verified. The optimized parameter set with the lowest bias will be selected for further optimization of the simulation of the cyclic deformation behavior. However, one key piece of information was not ignored in the optimization procedure, which is the yield stress at zero plastic strain. That parameter is equivalent to the yield point.

That said, when kinematic and isotropic hardening parameters are used, they do not yield equally. The major emphasis of this analysis is the high cycle fatigue range. The critical resolved shear stress should be significantly lower than the macroscopic tensile yield stress. However, which material characteristic calibration should be used? The usual technique for estimating fatigue limit requires a statistical S-N curve. Therefore, the critical resolved shear stress is used as an initial yield point instead of the macroscopic yield stress. The influence of using several back stresses with respect to the developed strain is seen in Figure 4.31. Figure 4.31 clearly depicts a stabilizing mechanism following loops of accumulation. Once some loops are complete, the strain amplitude decreases.

The 10 back stresses nearly have the same representation as to the 5 back stresses in Figure 4.32. Ten back stresses over-define the deformation behavior since it results in approximately similar loop characteristics as five back stresses even when the energy produced by the loops is higher for 10 back stresses. However, a higher

Table 4.6 Ten-pair hardening kernel function for AlSi10Mg and Ti-6Al-4V, respectively

AlSi10Mg	1st	2nd	3rd	4th	5th
A	369.74±17	355.58±14.1	280.42±9.02	465.40±17.31	437.40±11.87
B	80±0	90±0	140±0	70±0	80±0
k	19.57±0.75	27.39±0.82	29.61±1.06	43.74±1.24	49.60±1.13
Reduced Chi-sq	6.97	9.06	11.22	9.24	8.59
R-square	0.35	0.38	0.45	0.41	0.54
	6th	7th	8th	9th	10th
A	176.31±7.10	131.81±11.72	359.37±0.64	369.86±0.63	13.56±0.24
B	180±0	180±0	160±0	160±0	180±0
k	60.23±1.76	78.93±3.02	40±0	50±0	150.14±0.11
Reduced Chi-Sq	10.92	24.07	29.03	34.42	1.39
R-square	0.36	0.08	−0.27	−0.24	0.81

Ti-6Al-4V	1st	2nd	3rd	4th	5th
A	1159.30±14.50	982.02±20.61	2053.70±22.60	1277.00±17.80	969.10±81.90
B	400±0	200±0	160±0	160±0	160±0
k	40±0	200±0	200±0	300±0	368.87±6.48
Reduced Chi-sq	1186.45	835.75	862.38	646.72	584.31
R-square	−1.67	−0.08	−0.17	−0.08	0.23
	6th	7th	8th	9th	10th

(continued)

Table 4.6 (continued)

AlSi10Mg	1st	2nd	3rd	4th	5th
A	6266.20 ± 73.10	1690.60 ± 20.90	1632.60 ± 17.00	265.80 ± 4.12	148.23 ± 1.10
B	80 ± 0	130 ± 0	320 ± 0	750 ± 0	590 ± 0
k	250 ± 0	450 ± 0	400 ± 0	593.86 ± 2.41	699.41 ± 0.90
Reduced Chi-sq	3687.50	852.48	2992.38	110.40	6.58
R-square	-4.47	0.03	-2.87	0.89	0.98

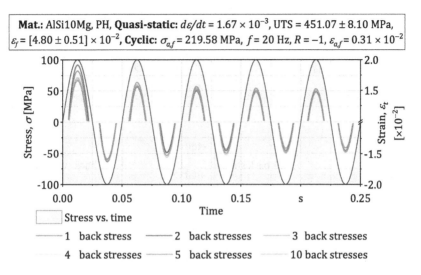

Mat.: AlSi10Mg, PH, **Quasi-static:** $d\varepsilon/dt$ = 1.67 × 10^{-3}, UTS = 451.07 ± 8.10 MPa, ε_f = [4.80 ± 0.51] × 10^{-2}, **Cyclic:** σ_{af} = 219.58 MPa, f = 20 Hz, R = −1, ε_{af} = 0.31 × 10^{-2}

Figure 4.31 Time versus stress-strain dependence under cyclic loading in AlSi10Mg at 100 MPa

number of back stresses simulates the ratcheting behavior much more progressively. In Figure 4.33, the stabilization process is also evident. But various back stress pairs are somewhat different since the change in strain response is less significant for Ti-6Al-4V, which means a smaller number of back stresses will be sufficient to simulate the ratcheting behavior in Ti-6Al-4V.

Figure 4.34 has 15 total loops simulated where loops stabilize visibly after the sixth loop has been completed. The small change in strain amplitude after this is negligible. The simulation with one back stress is significantly soft in the first loop but hardens at a considerably faster rate in the subsequent loops. So higher number of back stresses led to more accurate than progressive cyclic hardening. A similar 15 loop analysis was performed for Ti-6Al-4V, and the best fitting result will be utilized later in the simulation of cyclic deformation. The optimized back stress configuration is compared with the stable hysteresis loop from the experimental values. Worth mentioning that the deformations simulated in Figure 4.31, Figure 4.33 and Figure 4.34 are under consideration of kinematic hardening parameters only. This isolation is for the purpose of optimizing the isotropic and kinematic hardening parameters; everyone is on its own to minimize the number of experiments required for the calibration of the cyclic deformation simulation in a combined kinematic and isotropic setting.

Mat.: Ti-6Al-4V, PH, **Quasi-static:** $d\varepsilon/dt = 1.67 \times 10^{-3}$, UTS = 1280.59 ± 16.91 MPa, $\varepsilon_f = [11.99 \pm 1.67] \times 10^{-2}$, **Cyclic:** $\sigma_{a,f} = 844.48$ MPa, $f = 5$ Hz, $R = -1$, $\varepsilon_{a,f} = 0.75 \times 10^{-2}$

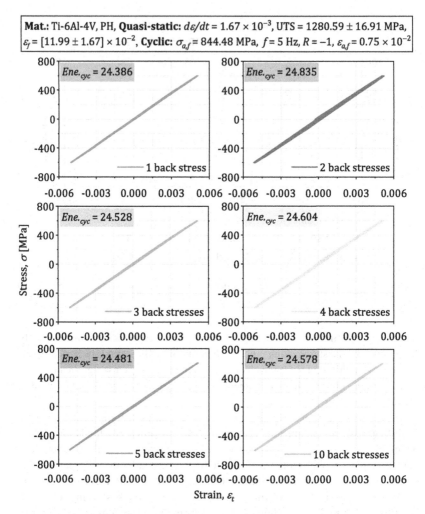

Figure 4.32 Hysteresis evolution optimization by increasing kinematic hardening pairs of AlSi10Mg at 100 MPa stress amplitude

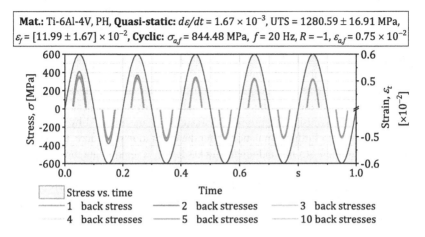

Figure 4.33   Time versus stress-strain dependence under cyclic loading in Ti-6Al-4V at 600 MPa stress amplitude

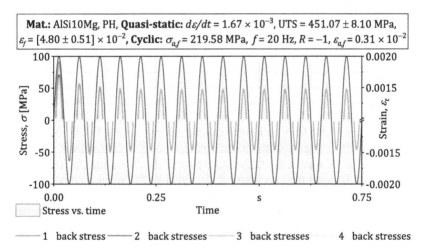

Figure 4.34   Time versus stress-strain dependence under cyclic loading in AlSi10Mg at 100 MPa for 15 loops of simulated loading

In Figure 4.35, the use of one back stress is ruled out as the hardening effect, and the shifting phenomena are not obvious. The preceding curve fitting stage is verified, as the stress softening setup cannot correctly simulate the hardening process. The cyclic hardening effect and the shifting phenomena are more discerned in the later back stresses. The five back stresses and 10 back stresses yield a fairly similar simulation result. In the AlSi10Mg simulation, this also occurred. After the five-back stresses are applied, the deformation has already been over-defined.

The back stress resistance primary characteristics are followed by subsequent simulation runs. In further simulations after the calibration procedure, the application setting of the isotropic and kinematic hardening is appropriate through the 2, 3, 4, 5-axis (back) stress simulations, which are employed to simulate stress magnitudes of 100, 120, and 140 MPa for AlSi10Mg, and 600 and 800 MPa for Ti-6Al-4V. The stabilized loops from the simulations are compared with the stabilized loops from the experimental results. Figure 4.36 indicates that four back stresses result in the optimum material calibration. Still, though, the simulation and experiment data points agree well in tension. Simulation and experimental data are not similarly accurate in agreement in compression. The simulation outcome is, however, showing enough accuracy to show important fatigue characteristics such as plastic strain amplitudes. Furthermore, experimental results had their own statistical variation, which reflects in the simulation. The author thus recommends future work to perform the calibration procedure on data with various strain rate levels or develop the applied cyclic plasticity model into a strain rate sensitive model in future work. Furthermore, the stress amplitude that is involved in the comparison in Figure 4.36 is already in the HCF regime, where strain measurements are somewhat affected by the fact that the proportions of deformation in this regime are relatively very small.

In Figure 4.37, the three back stress calibrations best suit the Ti-6Al-4V experimental data. The influence of strain rate is not present in this comparison since the tests of Ti-6Al-4V were done at 5 Hz of frequency. The discrepancy for Ti-6Al-4V is much less between the experiment and simulation. In the simulation, the simulation representation is nearly identical to the experiment result. The domain of compressive stress-strain shows a greater convergence from the AlSi10Mg result comparison. This is accredited to AlSi10Mg's higher degree of nonlinearity under dynamic loading conditions. The calibration procedure for Ti-6Al-4V yields a considerably better outcome compared to the calibration procedure for AlSi10Mg. The nonlinear stress-strain behavior of Ti-6Al-4V only exists in the upper range of the LIT, which is in the LCF regime. Since this transition area spans the range of a wider range of stress amplitudes, the calibration of AlSi10Mg is not ideal and suffers from fitting inaccuracies, especially in the compression portion of the hysteresis loop.

Mat.: Ti-6Al-4V, PH, **Quasi-static:** $d\varepsilon/dt$ = 1.67 × 10^{-3}, UTS = 1280.59 ± 16.91 MPa, ε_f = [11.99 ± 1.67] × 10^{-2}, **Cyclic:** σ_{af} = 844.48 MPa, f = 5 Hz, R = −1, ε_{af} = 0.75 × 10^{-2}

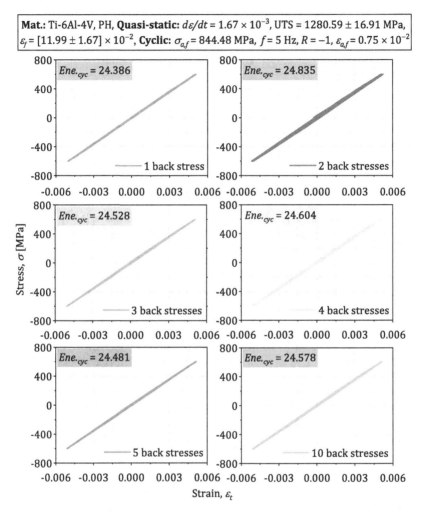

Figure 4.35 Hysteresis evolution optimization by increasing number of kinematic hardening pairs of Ti-6Al-4V at 600 MPa stress amplitude

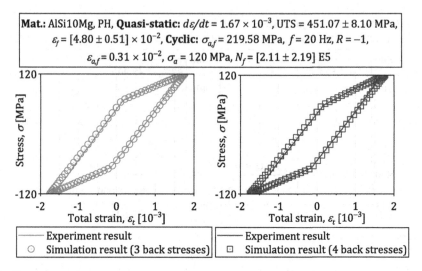

Figure 4.36 Comparison of experimental and simulation results of one hysteresis loop of AlSi10Mg at 120 MPa stress amplitude with 3 and 4 back stresses, respectively

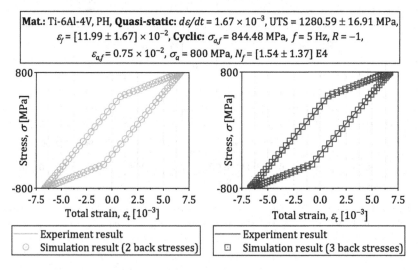

Figure 4.37 Comparison of experimental and simulation results of one hysteresis loop of Ti-6Al-4V at 800 MPa stress amplitude with 2 and 3 back stresses, respectively

Segmentation and discretization of cyclic stress-strain response in the LIT were shown to be effective in producing load range-specific optimization of back stress pairs in AlSi10Mg and Ti-6Al-4V. There are 3 distinct portions for the AlSi10Mg, and there are 2 distinct parts for the Ti-6Al-4V. The applied curve fitting functions will compensate the transition regions, which are not specified as a particular section due to the high nonlinearity of the material behavior in these regions. Here, the author concludes that the number of sections of the material's cyclic stress-strain response plus one is the optimal number of back stresses. One back stress cannot offer a near to the experiment data material calibration, but just a very rough estimation of the material's cyclic characteristics. Having an over-defined material calibration is not more accurate and can lead to a nonefficient simulation. Here, a limited number of experiments were utilized, yet the calibration was achieved by a numerical twin using Abaqus software. The experimental stress-strain response results were first organized based on the estimated slopes of the curves. An intriguing effect appears in the comparison of experiment and simulation data simulation. The cause of this outcome is perhaps connected to the statistical character of material properties.

Mesh sensitivity

The quality of simulation results is significantly influenced by mesh size, especially when the as-built defects from μ-CT are included in the simulations. Finer meshes produce better results but exponentially increase the expense of calculation in terms of hardware and time. As a result, the mesh sensitivity test must be performed first to select the optimum mesh capable of producing results that are equivalent to experimental results in an acceptable amount of time. The mesh sensitivity research is based on a direct cyclic examination of a Ti-6Al-4V porous model from μ-CT under a stress amplitude of 800 MPa with $R = -1$. The total strain amplitude achieved during a stable cycle is the argument that will be considered and optimized for in this sensitivity analysis. The goal of mesh sensitivity research is to find the right number of elements to give outcomes that are equivalent to experimental values while reducing computation time. This, for instance, is different from the established concept of mesh sensitivity analysis that dictates reaching a stable value of stresses at notches. The roots of the notches at the defects are considered anyway to be locations of singularity, and the values of stresses computed at these locations are just analyzed qualitatively since the numerical results reached in the simulations might be unreliable due to inhomogeneities of the macro and mesoscales. The global seed (G_s) distribution, which specifies the number of nodes and elements, is the variable in this study. From $G_s = 0.5$, which produced coarse elements to $G_s = 0.07$, which produced the finest element, eight different values of global seeds were examined. Table 4.7 shows that decreasing G_s, which represents a decrease in the distance between two vertices, increases the number of elements exponentially. Adversely, the time it takes for computation grows exponentially. The stabilized result took Abaqus solver 82 minutes to compute in the case of $G_s = 0.4$, and 1824 minutes in the case of $G_s = 0.07$ (30.5 hours). After evaluating several values of G_s, we can finally select $G_s = 0.1$. for future simulations. Figure 4.38 shows that the greatest total strain attained with $G_s = 0.1$ is $6.70 \cdot 10^{-3}$. This is the saturation limit, which specifies the greatest overall strain that this model can achieve in comparison to an acceptable experimental error based on the experimental results. Beyond this stage, the number of mesh elements has no influence, and the calibration and modeling parameters are alone in influence. G_s will be adjusted to 0.1 for subsequent simulations. It delivers accuracy within 99.82% of the saturation limit at $G_s = 0.1$ while keeping the computational cost under control.

Table 4.7 Function of the influence of global seed dimension G_s on the total strain amplitude

G_s	0.5	0.45	0.40	0.30	0.20	0.10	0.08	0.07
No. Elems. E3	196	210	229	257	296	331	590	716

Further simulations were carried out under stress control, which means that the maximum total strain achieved at each cycle decreases with time until saturation is reached since the applied model is more hardening-dominated than softening. This will be analyzed by isolating the influence of one back stress through programing a UMAT, which is to be used in Abaqus that uses one back stress for hardening based on the Armstrong-Frederick (AF) hardening rule. Also, isotropic hardening, which states that a greater stress level is required to obtain the same total strain in a subsequent cycle, is applied in this user-defined law. The Bauschinger effect causes the minimum total strain to deviate from the maximum total strain on the other side of the hysteresis. Both simulations tests under 600 MPa (Figure 4.39) and 800 MPa (Figure 4.40) reveal these two findings for the first three cycles. A log scale is used to visualize the hardening influence since the applied stresses are already in the late LCF to initial HCF range which is characterized by low strains and minute hardening behavior as well. Furthermore, cyclic hardening causes a decrease in overall strain. There are two components to the overall strain, which are represented as

$$\varepsilon_t = \varepsilon^{pl} + \varepsilon^{el} \qquad (4.74)$$

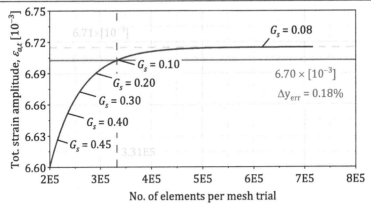

Figure 4.38 Analysis of the influence of geometric nonlinearities and mesh refinement on the strain state [304]

where the subscript t denotes total strain, the superscript pl denotes constitutive plastic strain, and the subscript el denotes constitutive elastic strain. The material is first in a virgin state (i.e., no prior deformation has occurred). Isotropic hardening contributes to the development of plasticity until it reaches saturation. Kinematic hardening, on the other hand, continues to develop and accumulate even after saturation.

As a result, kinematic hardening plays a significant role in the development and build-up of creep plasticity beyond saturation till failure. This phenomenon is seen under symmetrical loading ($\sigma_m = 0$) as well as asymmetrical loading ($\sigma_m \neq 0$) as an outcome of the Bauschinger effect, especially at notch roots of defects at which the micro stresses are not corresponding in load ratio to the applied remote amplitude. The cyclic response of 800 MPa exhibits the same phenomenon in Figure 4.40. However, in comparison to the 600 MPa test in Figure 4.39, the amount of reduction in the total strain at saturation is greater; this is because, at 800 MPa, more plastic strain occurs in each cycle, especially in the first cycles, which correspond well with the constant amplitude tests of section 3.3.4 for Ti-6Al-4V. The critical resolved shear stress of Ti-6Al-4V was estimated to be 450 MPa [325]. Reasonably enough, the plasticity induced by 600 MPa will be smaller than that induced by 800 MPa.

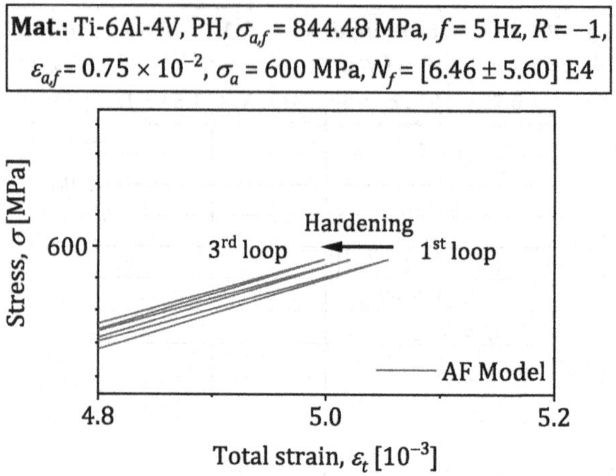

Mat.: Ti-6Al-4V, PH, $\sigma_{a,f} = 844.48$ MPa, $f = 5$ Hz, $R = -1$, $\varepsilon_{a,f} = 0.75 \times 10^{-2}$, $\sigma_a = 600$ MPa, $N_f = [6.46 \pm 5.60]$ E4

Figure 4.39   Three $(+/+)$ hysteresis tips of Ti-6Al-4V at 600 MPa of stress amplitude with a stress ratio of $R = -1$

Plastic strain can be calculated using either the elastoplastic stiffness matrix E_{elpl}, or strain decomposition, as shown in Equation 4.74. In general, simulation findings deviate from experimental results by a factor of 1% in the case of the 800 MPa test and 2% in the case of the 600 MPa test, which is computed as

$$\%ERROR = \left| \frac{\varepsilon_{a,t}^{exp.} - \varepsilon_{a,t}^{sim.}}{\varepsilon_{a,t}^{exp.}} \right| \cdot 100 \qquad (4.75)$$

The stabilized hysteresis loop was used to compare experimental and simulation findings. At the 10,000th cycle, the experimentally stabilized loops were extracted. The simulation results were based on the saturation of the Fourier term inside the direct cyclic analysis in Abaqus, and based on that, a stress-strain representation can be plotted. The maximum total strain reaches a saturation level before continuing to increase until failure. The greater deviation factor in the 600 MPa test is related to the AF model's performance. At greater strain ranges, the AF model, as well as other models based on the AF model, such as Chaboche and Ohno-Wang, tend to perform better $\varepsilon_t \geq 0.5\%$ [326]. The load range-related back stress concept has evolved in this thesis because of this. Back stress increases exponentially in the AF model, with a small variation throughout the knee section and a low strain range. As

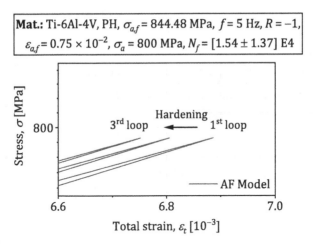

Mat.: Ti-6Al-4V, PH, $\sigma_{a,f}$ = 844.48 MPa, f = 5 Hz, R = −1,
$\varepsilon_{a,f}$ = 0.75 × 10⁻² , σ_a = 800 MPa, N_f = [1.54 ± 1.37] E4

Figure 4.40 Three (+/+) hysteresis tips of Ti-6Al-4V at 800 MPa of stress amplitude with a stress ratio of $R = -1$

a result, at higher strain ranges, the AF model can forecast a superior quality total strain. In the macroscopic elastic range, the experimental and simulation values, 600 and 800 MPa are comparable; however, the difference in the amount of plastic strain is higher. The continuum model parameters are to blame for this because the microstructure of as-built SLM Ti-6Al-4V is so inhomogeneous and highly variable from one location to another within a specimen. We were only able to link the microstructure of as-built SLM Ti-6Al-4V to the AF model using continuum model parameters, in this case, b, Q, C, and γ. However, microstructure sensitive crystal plasticity finite element calibration of the continuum plasticity models is highly recommended to achieve more versatile results [124, 125]. When comparing continuous hysteresis loops for AlSi10Mg, such as the one shown in Figure 4.41, the strain energy density decreases (i.e., the area bounded by the hysteresis loop) as the number of loops increases.

The capacity of an alloy's materials to resist greater load reversals, which impacts the material's fatigue limit, is known as its plastic damage tolerance. Plastic damage continues to build after saturation is attained because of yield surface translation in the three-dimensional stress field. Figure 4.41 shows cyclic hardening, which is indicated by increased stress levels at two regulated strain levels.

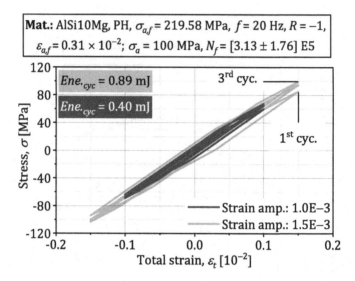

Figure 4.41 Simultaneous representation of hysteresis translation and expansion for AlSi10Mg at two different strain amplitudes [304]

4.4.3   Multiaxial Deformation Sensitivity

Besides correctly reproducing the cyclic deformation behavior as closely as feasible to the real scenario, the combined hardening material property has been established, and lifetime quantification based on shear stresses is explained in section 4.5. Simulations can now be performed to investigate the stress state of internal flaws. Fatigue life prediction under multiaxial stresses has been studied for a long time, and as previously discussed in section 2.1, voids in additive manufacturing components are highly challenging to avoid. Due to such flaws, the component's uniaxial load will be converted into a multiaxial load at the internal defects. Furthermore, because the practically usable fatigue life is in the high cycle fatigue region, the major study subject is multiaxial high cycle fatigue life prediction. A variety of multiaxial loading fatigue life prediction models have been developed. Most of the models, however, are focused on the low cycle fatigue area. Thus, research is being conducted to see whether such a technique is capable of predicting high cycle fatigue life under multiaxial conditions [327]. There are two sorts of models that are commonly employed in this circumstance. Equivalent stress models and critical plane models are the first and second types, respectively. The comparable stress model is a static yield criterion extension. To forecast the comparable fatigue life, the equivalent stress history will be utilized in conjunction with uniaxial stress-life data [328, 329].

Figure 4.42 shows the step-by-step method of transferring a μ-CT scan of a Ti-6Al-4V specimen to a finite element mesh. Figure 4.42c magnifies a cluster of subsurface pores, revealing a shrinkage pore typical of SLM and cast alloys caused by the collapse of gas pressure during cooling [197]. The mesh is presented in Figure 4.42e with $G_s = 0.1$. Two smaller pores at twelve and six o'clock suggest a potential location of crack coalescence under cyclic stress, in addition to the main flaws. Furthermore, a cyclic stress amplitude of 800 MPa is used to demonstrate the stress distribution around the defect; nevertheless, the internal stress on the defect is more than 450% of the applied distant stress amplitude. Following the conventions of macroscale severe plastic deformation, this should lead to a failure since this stress level surpasses the macroscopic yield of the material three-fold. However, the model is calibrated for fatigue loading in the HCF regime, and the resulting stress values are qualitative and not quantitative of the localization of strain under the remote fatigue loads.

The stress distribution legend highlights the possible fracture development directions from this defect qualitatively. Microstructural damage at these sites has a significant impact on the early fracture development phase. A representative volume element is suggested in a later section in the debate to capture such tiny microstructural characteristics while still obtaining a viable converging FE model respecting the convention of crystal plasticity, which is more suitable at this scale. The constitutive parameters for continuum or crystal plasticity might be homogenized and improved even further by using an optimization approach to adjust them [330], resulting in a superior elastoplastic response prediction. The collection of material parameters, on the other hand, cannot be generally used for various batch configurations with various thermal histories until a multiscale microstructurally sensitive setup is used. A qualitative evaluation of the influence of multiaxial stress can still be carried out by numerous approaches, such as Goodman, Gerber, and Soderberg's relations, which are based on this, with the concept of von Mises equivalent stress or hydrostatic stress being used to determine the effective stresses [331]. As the connection between stress amplitude and fatigue life is changed by a multiaxiality factor, von Mises equivalent stress amplitude with multiaxiality factor adjustment is also employed [332]. The modified Manson-McKnight model was also utilized. After then, the comparable stress is specified in terms of the "pseudo stress" range [333]. The critical plane model is the second kind. Fatigue fractures frequently nucleate on key shear planes, which motivated the development of such models. Normal stresses cause

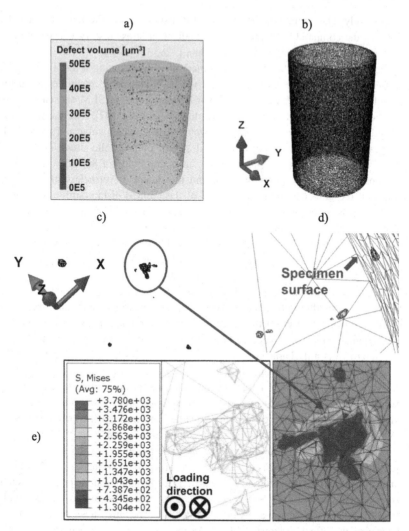

Figure 4.42 Application of stress amplitude at 800 MPa in Ti-6Al-4V: a) 3D defect volume distribution; b) 3D surface tessellation of the scanned volume; c) clusters of defects; d) multiple defects at different depth levels from the surface that respect the same triad in (c); e) stress concentration at a cluster of defects [304]

dislocation along slip lines by opening the fracture, reducing friction between crack surfaces, and causing shear stress. As a result, critical plane models usually employ a mix of normal and shear stress on the crucial plane [65, 329]. In this section, we will evaluate stress triaxiality at an internal defect as a measure of the possibility of failure at a respective point. This has the advantage of quantification of the proportion of shear inside a stress tensor. The stress triaxiality factor estimation is shown based on the formula [271, 334]

$$\frac{\sigma_h}{\sigma_{eq}} = \frac{\frac{1}{3}(\sigma_{11}+\sigma_{22}+\sigma_{33})}{\sqrt{\frac{(\sigma_{11}-\sigma_{22})^2+(\sigma_{22}-\sigma_{33})^2+(\sigma_{33}-\sigma_{11})^2+6(\sigma_{12}^2+\sigma_{23}^2+\sigma_{31}^2)}{2}}} \quad (4.76)$$

where σ_h is the hydrostatic stress and σ_{eq} is the von Mises stress. In Figure 4.43 the evolution of T.F. is presented for the notch root of an internal defect where the highest strain localization is developed under simulated stress amplitudes of 600 and 800 MPa, respectively, for 20 cycles. The T.F. is also projected on the stress-strain developed at the same node.

Before discussing T.F., it is worth mentioning that although applied stress amplitude in the simulation is completely reversible hence $R = -1$, the developed stress-strain response is not symmetric around the origin. The cyclic hysteresis does not pass through the origin during unloading, indicating the development of mean strain. In the case of the 600 MPa stress amplitude, the developed maximum tensile stress was ∼ 1200 MPa, while for the remote stress amplitude of 800 MPa, the maximum tensile stress was ∼ 1600 MPa. Hence this defect developed a stress concentration factor of 2 in addition to a mean strain indicating cyclic creep. However, the plastic strain amplitude of this hysteresis loop is minimal and negligible in both cases, which verifies and qualifies the use of LEFM in modeling fatigue strength-based crack propagation from defects in Paris law. On the other hand, T.F. peaks in both cases in the first tensile loading cycle, indicating the highest amount of hydrostatic stress, which can prove that damage at such defects initiates a brittle cleavage fracture at a very early stage and can propagate or get retarded based on the rules of short crack propagation depending on the material's ductility. Therefore, the author suggests a damage initiation criterion from pores based on the forming limit diagrams (FLD) and T.F. to be developed in future investigations.

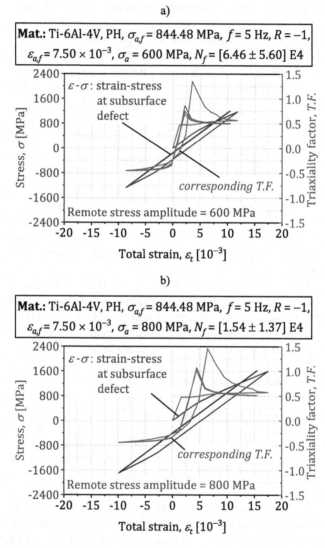

Figure 4.43 Stress triaxiality factor evolution at a subsurface defect for uniaxial remotely applied stress: a) 600 MPa; b) 800 MPa

Prospect of lower scale validation methods

The AF model, as well as Chaboche, predict a greater total strain which is due to microstructural deformation characteristics and strengthening mechanisms, which prevent dislocation movement and lead to hardening. Because dislocation movement is hampered, the effective slip length is shortened, resulting in a reduced total strain [335]. In the case of Ti-6Al-4V, this is a result of the ultrafine α'-Ti microstructure, as evidenced by the reduced elongation at fracture in a monotonic tensile test and early failure of as-built SLM Ti-6Al-4V under cyclic loading. However, utilizing crystal plasticity models and microscale modeling based on electron backscattered diffraction (EBSD) imaging method, this impact can be considered and accounted for. The crystal plasticity constitutive laws are recognized as valid laws at the scale of the microstructure around pores that were presented in Figure 4.42. By application of crystal slipping models [126], converging damage parameters at low remote stresses typical of very high-cycle fatigue are possible to obtain

$$\overline{\boldsymbol{L}}^{p} = \sum_{\alpha} \dot{\gamma}^{\alpha} \left(\overline{\boldsymbol{s}}^{\alpha} \otimes \overline{\boldsymbol{n}}^{\alpha} \right) = \dot{\boldsymbol{F}}^{p} \boldsymbol{F}^{p-1} \qquad (4.77)$$

where $\dot{\gamma}^{\alpha}$ is the slip system shearing rate, and $\overline{\boldsymbol{s}}^{\alpha}$ and $\overline{\boldsymbol{n}}^{\alpha}$ are unit vectors in the slip direction and slip normal plane direction, respectively, for the αth slip system. The viscoplastic flow rule for the shearing rate on the αth slip system is given by

$$\dot{\gamma}^{\alpha} = \dot{\gamma}_0 \left\langle \frac{|\tau^{\alpha} - \chi^{\alpha}| - K^{\alpha}}{D^{\alpha}} \right\rangle^{m} sgn\left(\tau^{\alpha} - \chi^{\alpha} \right) \qquad (4.78)$$

where $\dot{\gamma}_0$ is the reference shear rate, τ^{α} is the resolved shear stress, χ^{α} is the crystal back stress, K^{α} is the length scale-dependent threshold stress, D^{α} is the drag stress, and m is the flow exponent. The crystal slipping constitutive laws are widely applied to statistical and representative volume elements to obtain crack initiation lifetimes [126, 336].

In Figure 4.44, representative volume elements (RVE) of the investigated AlSi10Mg and Ti-6Al-4V are shown, which represent the simulation field for the understanding of the crack initiation process under cyclic loading in a very high-cycle fatigue regime for future investigation prospects. On this scale [337], attractive contributions were made based on the segmentation of single grains in the microstructure and were able to calculate crack initiation lifetime based on Tanaka-Mura dislocation gliding theory. Such micromechanical simulations were shown effective at enhancing the precision of model calibration on the macro scale,

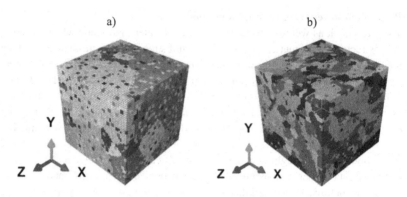

Figure 4.44 Three-dimensional models of the volume elements of investigated alloys: a) AlSi10Mg; b) Ti-6Al-4V [304]

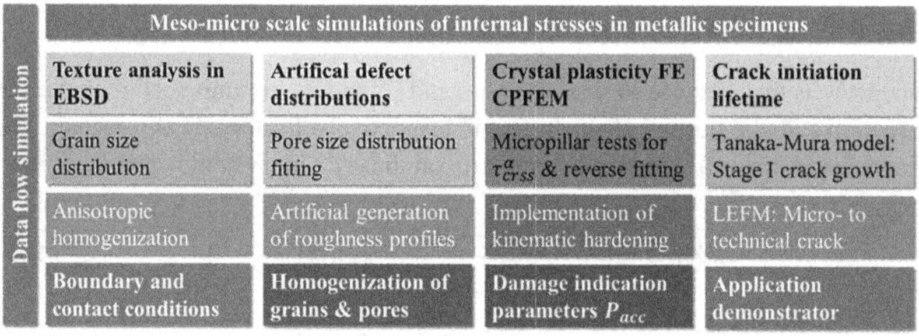

Meso-micro scale simulations of internal stresses in metallic specimens			
Texture analysis in EBSD	Artifical defect distributions	Crystal plasticity FE CPFEM	Crack initiation lifetime
Grain size distribution	Pore size distribution fitting	Micropillar tests for τ_{crss}^{α} & reverse fitting	Tanaka-Mura model: Stage I crack growth
Anisotropic homogenization	Artificial generation of roughness profiles	Implementation of kinematic hardening	LEFM: Micro- to technical crack
Boundary and contact conditions	Homogenization of grains & pores	Damage indication parameters P_{acc}	Application demonstrator

(Left vertical label: Data flow simulation)

Figure 4.45 Flow chart of the suggested simulation scheme on the meso-microscale

as shown in the literature [127]. The limitation of the macro-mesoscale methodology applied here initiated this ongoing discussion about meso-microscale effects where crystal sliding constitutive laws do not encounter the convergence problems of AF or Chaboche based cyclic plasticity models. In Figure 4.45, a concept is presented for meso-microscale crack initiation time analysis in which some elements were already presented in the literature, as discussed above. The scheme makes use of crystallographic orientation data on various deposition planes to capture the anisotropy of additively manufactured materials. In these three-dimensional representative volume elements, real-morphology pores from μ-CT are integrated with relevant surface roughness profiles from macroscale specimens. Consequently, the crystal plasticity finite element method (CPFEM) is applied to capture fatigue damage parameters based on critical resolved shear stress and crystallographic kinematic hardening. On this foundation, dislocation-based crystal damage criteria should be used to identify and advance stage I crack development, after which service components can be subjected to linear elastic fracture mechanics (LEFM) to determine the remaining fatigue lifetime. Unforced fatigue cracks that originate from slip bands on the maximum shear planes are likely to adopt a mode II shearing failure that will branch into mode I tensile failure depending on homogeneities encountered along the crack's course, according to Fatemi et al. [65, 329]. Awd et al. [304], for SLM AlSi10Mg, observed the phenomena in interrupted VHCF fatigue testing, where shear generated fractures branched and evolved into additional modes dependent on local microstructure and porosity as well as local loading/fracture variations at various stages in a specimen's life.

4.5   Shear-based Fatigue Damage Quantification

Additive manufacturing is frequently commended for its capacity to produce high-complexity components that create complex multiaxial damage under service loading. Although the fatigue experiments and data presented in this study are uniaxial, the analytical technique discussed in this section has all the major aspects required to undertake a multiaxial stress analysis of durability. The use of 3D μ-CT data, FE model, cyclic plasticity material law, and Fatemi-Socie damage parameter for complicated AM geometries that are expected to undergo multiaxial stress states are brought here to advantageous handling. More life prediction methods based on the concept of critical shear planes exist. For example, maximum normal stress amplitude was corrected by Smith, Watson, and Topper. The fatigue life is a function of the critical plane's maximum normal strain amplitude and maximum normal stress [338]. Furthermore, the Goodman, Gerber, and Soderberg criteria use the same mean stress as before, but shear stress components replace normal stress [331]. The Kandil, Brown, and Miller parameter considers the maximal alternation shear strain to be the primary damage component on the plane, with the normal strain serving as a backup [339]. The maximum alternating shear stress is considered the major damage component on a plane, whereas the maximum normal stress is considered the secondary damage component [340]. Findley parameter is comparable to the McDiarmid parameter. However, in terms of the material's ultimate strength and fatigue in torsion, an adjustable factor is used [341]. The maximum alternating shear strain is considered the major damage component by Fatemi, Socie, and Kurath, whereas the maximum normal stress is considered the secondary damage component [328]. The Chu, Conle, and Bonnen parameter is an energy-based model that is calculated by multiplying the shear and normal strain amplitudes on a plane with the corresponding shear and normal maximum stresses on that plane. To optimize the outcome, an adjustable parameter is utilized [342]. The Glinka, Wang, and Plumtree parameter, on the other hand, is an energy-based parameter. The maximum shear and normal stress on that plane are connected to the shear stress and strain amplitude product [256]. As stated earlier, the majority of the models previously discussed are utilized in LCF fatigue life prediction; however, the adjustable factor in each model will be modified to accommodate for changes in the forecast of the lifetime. The Fatemi-Socie-Kurath model produced a satisfactory result in HCF multiaxial fatigue life prediction in an earlier study [312, 328]. Moreover, it can be connected to homogenized state variables resulting from the FE simulations. To minimize the difference between experimental and simulation findings, an optimization method is required. This is seen in the work of [326, 343, 344], which employed a variety of metrics to better assess ratcheting, mean stress relation, and cyclic hardening. Genetic porosities and surface roughness also have a role in the elastoplastic response. Consequently, by simplification of the μ-CT scan, a crack propagation simulation and fatigue lifetime analysis can be conducted, such as in section 4.3. This result will be used to calibrate the Fatemi-Socie model on the macroscale to calculate synthetic Woehler curves.

4.5.1   Fatemi-Socie Planar Fatigue Damage Model

The Fatemi-Socie (FS) life prediction model was developed using a modified version of Brown and Miller's critical plane method [345]. The change is necessary because Brown and Miller's critical plane method was not implemented to include load-specific fatigue lifetime [346]. The critical plane model is used to show that localized plastic deformation in persistent slip bands is implicated in the onset of fatigue cracks. The continuous slip band direction is roughly matched with the greatest shear strain direction, according to prior observations [347]. Furthermore, under various stress conditions, the fatigue fracture begins on the greatest shear plane [345, 347]. All these findings demonstrate the strong link between fatigue onset and maximum shear strain. The Brown and Miller equation's normal strain term is translated to normal stress, and the relation can be formed [345]

$$\gamma_{max}\left(1 + \frac{k\sigma_n^{max}}{\sigma_y}\right) = constant \tag{4.79}$$

The k is a constant that had been calibrated by fitting uniaxial data to pure torsion data, and it is known as the Fatemi-Socie damage parameter, σ_n^{max} is the maximum normal stress, γ_{max} is the maximum shear strain. Because the strain is dimensionless, the maximum normal stress is normalized by the yield stress. To adjust for different forms of fatigue life, the value of σ_y will be alternated based on the corresponding back stress value. The maximum shear strain is affected by the maximum normal stress in this life prediction technique which can be deduced from FE simulations [348], accordingly

$$\frac{\Delta\gamma_{max}}{2}\left(1 + k\frac{\sigma_n^{max}}{\sigma_y}\right) = \gamma_f'\left(2N_f\right)^{c_0} + \frac{\tau_f'}{G}\left(2N_f\right)^{b_0} \tag{4.80}$$

The k value is critical for accurate life prediction, γ_f' is torsional fatigue ductility coefficient, c_0 is the torsional fatigue ductility exponent, τ_f' is torsional fatigue strength coefficient, b_0 is the torsional fatigue strength exponent, and G is the shear modulus. Although it can be computed without the axial and torsional fatigue, the unique properties of the model can be fitted [349]

$$k = \left[\frac{\gamma_f'\left(2N_f\right)^{c_0} + \frac{\tau_f'}{G}\left(2N_f\right)^{b_0}}{(1 + v^e)\frac{\sigma_f'}{E}\left(2N_f\right)^{b} + (1 + v^p)\varepsilon_f'\left(2N_f\right)^{c}} - 1\right]\frac{2\sigma_y}{\sigma_f'\left(2N_f\right)^{b}} \tag{4.81}$$

where v^e is the elastic Poisson's ratio, v^p is the plastic Poisson's ratio, σ_f' is the axial fatigue strength coefficient, b is the axial fatigue strength exponent, ε_f' is the axial fatigue ductility coefficient, c is the axial fatigue ductility exponent. Such a computation, on the other hand, necessitates a large number of variables. The fatigue lifetime estimate will be more difficult to calculate in terms of shear components. According to prior research, k is a material property that is determined by both uniaxial and multiaxial data. Due to a modification in the projection of long cycle life, however [328], in this study, it will be treated as a load level-dependent parameter. The fitting and calibration of k value lead to the life forecast calculation. Although the damage parameter in the Fatemi-Socie model refers to the material's torsional characteristic rather than its uniaxial characteristic, a set of equations for mapping the uniaxial to torsional properties will be applied. Hence, no additional experimentation is required. The von Mises criteria can be used to convert uniaxial characteristics to torsional properties using the following equations

$$\tau'_f = \frac{\sigma'_f}{\sqrt{3}}; \gamma'_f = \sqrt{3}\varepsilon'_f; G = \frac{E}{2(1 + v^e)}; b \approx b_0; c \approx c_0 \qquad (4.82)$$

All the elements in Equation 4.80 can be obtained by curve fitting or simple calculation. k is considered here as a material property that varies depending on the material and load situation. The maximal shear strain, however, is the most critical component that will calibrate the model from the FE simulations. The maximum shear strain will be calculated as the loading process, and the stable response will be simulated in Abaqus. Curve fitting will be used to compute or get all additional material parameters. Equation 4.82 describes the property mapping between torsional and uniaxial properties. The k can be computed from the reference using a highly complex relationship, which needs more experimentation to get all of the needed variables. The value of k is the only variable that is unknown at this point which is calibrated by the calculated fatigue lifetimes in section 4.3. The goal of this method is to compute the k value backward and then link it to a parameter, which is a damage parameter, that is related to the material and load level. After that, the relationship is calibrated only on the basis of the values acquired in the elastoplastic cyclic deformation simulation. The estimated k will then be applied to the Fatemi-Socie model, and if there is a significant discrepancy, it will be recalibrated in an iterative manner.

AlSi10Mg / MPa	k-value	**Ti-6Al-4V / MPa**	k-value
100 MPa	2.40	**600 MPa**	9.00
120 MPa	1.75	**800 MPa**	4.35
140 MPa	1.30		

Table 4.8 Resulting k-value based on the first estimate for AlSi10Mg and Ti-6Al-4V

The first step is to use a reference to determine the cycle to failure for various materials under various stress amplitudes. The value of k for both materials and different loading amplitude will be calculated based on LEFM simulations [32, 50]. The calculated values are presented in Table 4.8 based on Equation 4.81. The computed k value will be linked to a damage parameter derived through model simulation data. Because the whole experimental and simulation data are stress-regulated, there is no point using stress amplitude across various loops. The strain amplitude is thus responsive to the development of cyclic deformation and damage. As the stress amplitude and material vary, these values change as well. The absolute strain amplitude, on the other hand, cannot offer adequate information since it does not describe the progression of damage as response saturates.

The absolute strain amplitude varies mainly depending on the material and stress amplitude. However, there is a distinction between the initial loop and the stabilized loop. The difference between the stabilized loop and the initial loop, on the other hand, offers an idea of what specific damage progression can be utilized for calibration. The ratio of stress amplitude difference of stable loop and first loop is generated by the isotropic hardening parameter calibration Equation 4.71, see Table 4.9. In this case, the reverse is true, but a simpler notion will be utilized. For each material in different stress amplitudes, a factor of the ratio between the strain amplitude for the first loop and stabilized is computed. This factor will be linked with the previous stage's factor k. This component was chosen because it provides a solid description of the material damage parameter based on the accumulated hardening in the material at various loading amplitudes. The simulation output will be used to determine the strain amplitudes. A clear linear connection can be established in the alloys AlSi10Mg and Ti-6Al-4V, see Figure 4.46. The slope and y-intercept values are shown in Table 4.10, Table 4.11 and Table 4.14.

Table 4.9 For both materials and varied stress amplitudes, the ratio of the first loop's and the stabilized loop's strain amplitude

AlSi10Mg / MPa	$\varepsilon_0/\varepsilon_S$	Ti-6Al-4V / MPa	$\varepsilon_0/\varepsilon_S$
100	1.14	600	1.04
120	1.16	800	1.05
140	1.16		

Table 4.10 Parameters of the functions of the Fatemi-Socie parameter k

	Slope	y-intercept
AlSi10Mg	−48.31	57.56
Ti-6Al-4V	−351.28	373.86

The slope and y-intercept values will be calibrated using the results of the simulation. In addition, because the k value is such a delicate factor, the value will be calibrated as iteratively as feasible. Slight variations (Table 4.12 and Table 4.15) will result in significant variance in the cycle to the failure calculation result; see Table 4.13 and Table 4.16.

As mentioned, a slight deviation in k value results in a great deviation in the result of the cycle to failure. The k value calibrated with Table 4.13 is not accurate enough. However, with further iterations of calculations, convergence to the correct values is achieved and presented in Table 4.15 and Table 4.16.

Table 4.11 Initial estimates of slope and y-intercept

	Slope	y-intercept
AlSi10Mg	−47.39	56.73
Ti-6Al-4V	−349.65	372.94

Table 4.12 The k value estimate based on the kernel functions of Table 4.11

AlSi10Mg / MPa	k-value	Ti-6Al-4V / MPa	k-value
100	2.61	600	9.78
120	1.99	800	5.15
140	1.53		

Table 4.13 Fatigue lifetime estimate optimized from the kernel function of FS parameter

AlSi10Mg / MPa	N_f	Ti-6Al-4V / MPa	N_f
100	59526	600	41679
120	24214	800	1522
140	13767		

Table 4.14 Recalibrated kernel function of the slope and intercept according to Equation 4.83

	Slope	y-intercept
AlSi10Mg	47.44	56.53
Ti-6Al-4V	351.41	373.99

$$FS = -\left(\left(G\gamma_f\varepsilon_{f-}\sigma_y\left(\frac{v_e + v_p}{2}\right)\right)\cdot\left(1 - \frac{b}{100}\right)\right) \qquad (4.83)$$

Table 4.15 Reoptimized k value based on the update of Table 4.14

AlSi10Mg / MPa	Recalibrated k-value	Ti-6Al-4V / MPa	Recalibrated k-value
100	2.35	600	9.01
120	1.72	800	4.35
140	1.27		

The highest in-plane shear strain generated under simulated cyclic loading determines the fatigue lifetime prediction model based on Fatemi-Socie damage in Equation 4.80. Based on simulation data of von Mises shear strain amplitude, the damage parameter k was calculated in an inverse iterative method. A

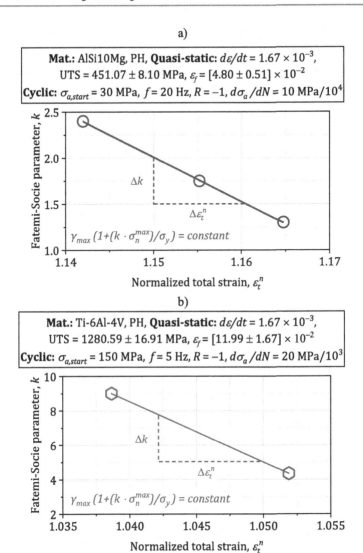

Figure 4.46 Relationship between Fatemi-Socie damage parameter k and strain ratio: a) AlSi10Mg; b) Ti-6Al-4V

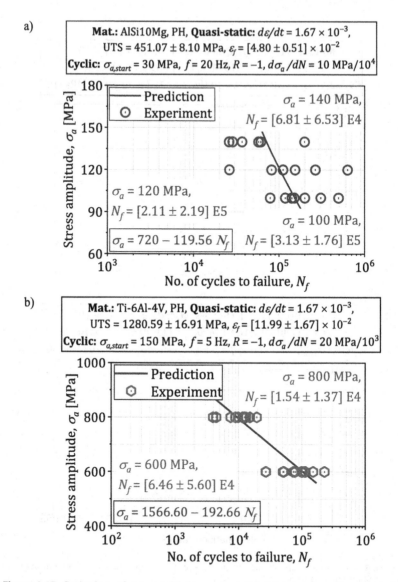

Figure 4.47 Projections of experental fatigue and predicted data were used to evaluate the proposed modeling scheme in low-cycle to high-cycle fatigue regimes: a) AlSi10Mg; b) Ti-6Al-4V

Table 4.16 Fatigue lifetime estimate reoptimized from the updated kernel function of Fatemi-Socie parameter

AlSi10Mg / MPa	N_f	Ti-6Al-4V / MPa	N_f
100	153502	600	104104
120	103399	800	9535
140	71278		

good match is observed in the load range when this simulation is performed in Figure 4.47, which shows a comparison of predicted and experimental fatigue data. The author, on the other hand, discusses the adaptation of this methodology in late high-cycle and very high-cycle fatigue in the next section, where the convergence of the employed plasticity model is more challenging due to small plastic strain increments. As a result, material laws on a smaller scale are expected to perform better in planned future studies. The calibrated k value in the first step produces a result that is not converging to the experiment value. The calibrating process based on LEFM and XFEM analysis from section 4.3 can produce a satisfactory outcome. Furthermore, a relationship between the k value and the ratio of strain amplitude between the first and stabilized loops enables to merge the influence of the Bauschinger effect.

4.5.2 Adaptation of the Model to VHCF Application

Certain components have a life expectancy of between 1E7 and 1E10 cycles in current technological applications. There are fewer experimental findings beyond 1E7 cycles that can be used to estimate the conventional fatigue limit using statistical methods. The S-N curve must be established in the gigacycle regime in order to ensure the true fatigue strength and safe life of mechanical components [350]. The technological importance of lightweight alloys such as AlSi10Mg and Ti-6Al-4V for aerospace and automotive industries is shown in section 3.1. The fatigue field regime of this typical application is VHCF [26]. However, VHCF lifetime prediction models are limited to microscale microstructure-sensitive crack propagation models without direct S-N relation such as Przybyla et al. [351] as well as prediction of S-N relations in the VHCF regime based on Baqsuin equation [352] in addition to dislocation-based models such as Tanaka-Mura [129]. This diversity of damage evolution mechanisms in VHCF has been discussed by Zimmermann [353] where the fact that VHCF is dominated by crack initiation is thoroughly presented. In addition, further researchers have built on the Fatemi-Socie (FS) damage parameter to predict the microstructure-sensitive fatigue behavior of the powder metallurgy (PM) Ni-based superalloy IN100 [354]. Furthermore, the shift from the surface to the subsurface was studied in terms of competing damage processes during the transition from the HCF to the VHCF regimes using the Fatemi-Socie parameter as a damage indication parameter [355]. The technical application of the concept extended to investigate the service life of a turbine blade that was subjected to multiaxial cyclic loadings. Premature blade failure was a result of high-cycle fatigue (HCF) mechanisms [356]. Therefore Fatemi-Socie parameter is valid to describe the *S-N* relationship in the VHCF regime since failure is dominated by localization of shearing that initiates cracks.

Gates and Fatemi [348] describe how the shear strain-life curve for the material that is to be analyzed is represented on the right-hand side of Equation 4.80. In the case that shear fatigue characteristics are not available for use in damage computation, the right side of this equation can be written in terms of uniaxial fatigue properties [357], which is equivalent to the right side here

$$
\frac{\Delta \gamma_{max}}{2} \left(1 + \frac{\sigma_{n,max}}{\sigma_y} \right)
$$

$$
= \left[\left(1 + v^e \right) \frac{\sigma_f'}{E} \left(2N_f \right)^b + \left(1 + v^p \right) \varepsilon_f' \left(2N_f \right)^c \right]
$$

$$
\cdot \left[1 + k \frac{\sigma_f'}{2\sigma_y} \left(2N_f \right)^b \right] \tag{4.84}
$$

where v^e is elastic Poisson's ratio, v^p is Poisson's ratio for fully plastic conditions (set for 0.5), and all other fatigue properties correspond to the fully reversed uniaxial stress-life equation. The normal stress term is multiplied by the shear strain range to obtain the damage-dependent life. Damage parameters that are based only on stress or strain terms are unable to accurately represent material constitutive behavior. That is important because it precludes the prediction of fatigue damage in circumstances where only a static axial tension can exist. Gates and Fatemi [348] made use of a Mohr's circle analysis to create a connection between the fully-reversed uniaxial loading and the stresses and strains on the maximum shear plane in the maximum shear plane. The damage parameter can then be expressed in terms of uniaxial loading utilizing the relationship models that have been established. The following life prediction curve for the modified FS parameter based on uniaxial strain-life characteristics is produced for the modified FS parameter according to the methodology described in section 4.5.1

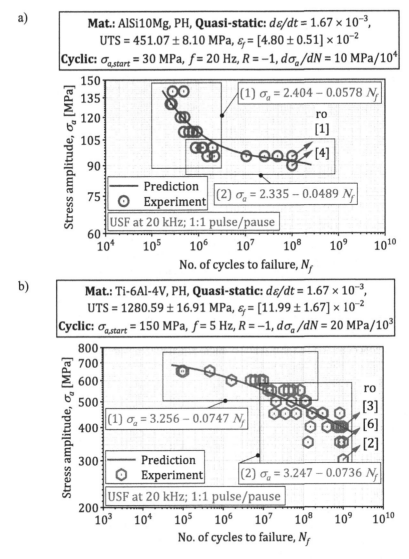

a)

Mat.: AlSi10Mg, PH, **Quasi-static:** $d\varepsilon/dt = 1.67 \times 10^{-3}$,
UTS = 451.07 ± 8.10 MPa, ε_f = [4.80 ± 0.51] × 10⁻²
Cyclic: $\sigma_{a,start}$ = 30 MPa, f = 20 Hz, R = −1, $d\sigma_a/dN$ = 10 MPa/10⁴

(1) $\sigma_a = 2.404 - 0.0578\ N_f$

ro
[1]
[4]

Prediction
Experiment
(2) $\sigma_a = 2.335 - 0.0489\ N_f$
USF at 20 kHz; 1:1 pulse/pause

b)

Mat.: Ti-6Al-4V, PH, **Quasi-static:** $d\varepsilon/dt = 1.67 \times 10^{-3}$,
UTS = 1280.59 ± 16.91 MPa, ε_f = [11.99 ± 1.67] × 10⁻²
Cyclic: $\sigma_{a,start}$ = 150 MPa, f = 5 Hz, R = −1, $d\sigma_a/dN$ = 20 MPa/10³

(1) $\sigma_a = 3.256 - 0.0747\ N_f$

ro
[3]
[6]
[2]

Prediction
Experiment
(2) $\sigma_a = 3.247 - 0.0736\ N_f$
USF at 20 kHz; 1:1 pulse/pause

Figure 4.48 Projections of experimental and predicted fatigue data were used to evaluate the proposed modeling scheme in high-cycle to very high-cycle fatigue regimes after adaptation: a) AlSi10Mg; b) Ti-6Al-4V

$$\frac{\Delta\gamma_{max}}{2}\left(1 + \frac{\sigma_{n,max}}{\sigma_y}\right)$$

$$= \left[\left(1 + v^e\right)\frac{\sigma'_f}{E}\left(2N_f\right)^b + \left(1 + v^p\right)\varepsilon'_f\left(2N_f\right)^c\right]$$

$$+ k\frac{\sigma'_f}{4G}\left(2N_f\right)^b \qquad\qquad (4.85)$$

The corresponding back stress to the applied load was used to calibrate Equation 4.85 based on the value of the fatigue strength coefficient σ'_f. This is in addition to the FS parameter calibration methodology described in section 4.5.1. In Figure 4.48, the comparison between experimental data and model results is projected, which shows good correspondence in producing synthetic S-N curves. This result can be reached provided the correct back stress value is applied at the stress amplitude at which the fatigue lifetime needs to be calculated. High cycle fatigue (HCF) of polycrystalline metals is concerned with the degree of heterogeneity of cyclic slip processes. The resistance to the cyclic slip process is proportional to the corresponding back stress, which validates the assumption on which the presented model is applied. In the HCF regime, the cyclic shear strain concentration factor inside the microstructure is greater than elastic loading [358]. The critical plane and the fatigue fracture plane are consequently related. The critical plane depends on both the stress state and the material characteristics in the FS model. The critical plane corresponds to the minimum hydrostatic stress damage since the damage is mainly induced by the deviatoric load that is quantified by the T.F. as discussed in section 4.4.3 [359]. Calculations relating to fatigue hot spots can then be used to determine the probability of fatigue fracture initiation in HCF and VHCF. From the distribution of state parameters or fatigue hot spots, one can deduce the distribution of fatigue lifetime. Due to the time and money involved in conducting statistically meaningful experimental research in the HCF-VHCF regimes, such inferences can have substantial utility. This exact concept will be applied in section 5.1. The driving factors for fatigue fracture development in a duplex Ti-6Al-4V alloy were explored utilizing that concept [360]. The microstructures with smaller relative primary grain sizes and lower volume percentages exhibit less variation and have smaller driving force magnitudes [361]. The driving factors were greatest at higher failure probability levels [362]. The concept is applied to calculate fatigue lifetimes and locations of fracture of Ti components of aerospace engines [363]. In adversity and at a submicron scale, the critical plane approach helped to determine the driving force for fatigue crack nucleation at twins in RR1000 superalloys [364]. Fatemi-Socie (FS) and Smith-Watson-Topper (SWT) were used to predict fatigue life on low C-Mn steel. The critical plane model resulted in an accurate assessment under different loading types [365]. However, the FS parameter predicted fatigue life poorly HCF regime in multiaxial fatigue tests on AZ61A magnesium alloy [366].

Bayesian Inferences of Fatigue-related Influences

5

The "industrial miracle" of Japan began in the mid-twentieth century. Much of the Japanese success has been ascribed to their management personnel's use of statistical tools and statistical thinking [367]. That had a long way to go in terms of improving the quality of manufacturing in other parts of the world. Fatigue is a random variable regardless of whether it is examined in the laboratory or in technological applications on-site. There is considerable literature that suggests a variation in the fatigue strength of identically shaped components/specimens made in the same technique [368]. Thus, statistical investigations of fatigue damage processes are required to ascertain their central tendency and dispersion around it.

Probability is a mathematical concept used to express uncertainty. The likelihood of a structure or component failing is a critical design parameter for structural analysis [369]. This value is often determined using material constants, the anticipated stresses to be applied to the component, a failure criterion (such as the von Mises or Tresca criteria), testing, and the selection of a probability model [370]. Baye's rule establishes a logical procedure for revising beliefs in the face of new knowledge. Bayesian inference is the process of inductive learning using Bayesian inference [371].

5.1 Phenomenological Statistical Learning

The use of mathematical and statistical models is critical in evaluating and forecasting the fatigue lifetime of machines and structures [372]. Consideration of the stress range, stress level, and size effect, in conjunction with efficient estimation

M. Mamduh Mustafa Awd, *Machine Learning Algorithm for Fatigue Fields in Additive Manufacturing*, Werkstofftechnische Berichte | Reports of Materials Science and Engineering, https:doi.org/10.1007/978-3-658-40237-2_5

of the related parameters, constitutes one of the most challenging and enticing problems that has not been completely resolved [370].

5.1.1   Extreme Value Statistics in Fatigue

This section introduces the extreme value probability distributions used in fatigue strength analysis, which are utilized consistently throughout the next parts and have a strong statistical foundation in correlating applied loading conditions to the expected service life, as well as some of their characteristics, particularly those related to lifetime issues [368]. This is studied under consideration of structural characteristics which stem from the processing route of the material or structure. The gamma distribution derives its name from the well-known gamma function, studied in many areas of mathematics [373]. Based on this distribution, the expression of Weibull and Gumbel-based fatigue strength will be introduced.

Extreme-value and stable probability density functions (PDFs)
The relationship between the gamma distribution and the exponential failure rate function allows the gamma functions and their derivatives to be used in fatigue analysis problems. The time span between load applications in certain laboratory or service conditions and the time to failure of specimens can be adequately described by the gamma distribution functions such as Weibull or Gumbel distributions [327, 374]. The gamma function is defined by the form

$$\Gamma(\alpha) = \int_0^\infty x^{\alpha-1} e^{-x} dx, \, for \, \alpha > 0 \qquad (5.1)$$

If the density function of the failure rate of X is given by Equation 5.2, then it has a gamma distribution with the parameters α and β and the definition [368]

$$f(x; \alpha, \beta) = \begin{cases} \frac{1}{\beta^\alpha \Gamma(\alpha)} x^{\alpha-1} e^{-x/\beta}, \, x > 0 \\ 0, \, elsewhere \end{cases} \qquad (5.2)$$

where $\alpha > 0$ and $\beta > 0$. The values of α and β change the central tendency and scatter of the distribution as shown in Figure 5.1 for the α parameter.

Figure 5.1 The dependency of central tendency bias on the gamma distribution α parameter

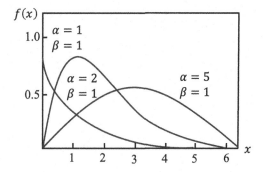

Additionally, the Weibull distribution is the main distribution to solve reliability and life-testing challenges, such as the time to failure or the duration of a component measured from some defined period until it fails. The Weibull distribution is intrinsically flexible in that it does not require the exponential distribution's lack of memory [130, 152, 375]. Creating predictive models of fatigue behavior is always a great challenge in the technology of additive manufacturing due to the immense number of factors that can be extremely complex to physically incorporate into a deterministic modeling scheme. It also became of great importance after adopting the technology in the aerospace industry since service conditions are as complex as manufacturing parameters in terms of deterministic description [11, 372]. Many companies tend to introduce additively manufactured parts into their airplanes; however, the qualification procedure is rigorous and tends to consume much more resources than the manufacturing itself. Aluminum alloys are mostly used as an engineering material for this purpose, followed by titanium alloys [2, 376]. Therefore, more research is being directed to understanding and predicting the fatigue behavior of such new parts based on statistical and artificially intelligent methods. In the next pages, it will be explained how the computation of predictive probabilistic percentile for the *S-N* curve of an additively manufactured specimen can be performed in a machine learning conceptual procedure that automatically learns the statistical model parameters without being explicitly programmed to do so. Two probability distribution functions were used in this thesis which is Weibull and Gumbel distributions. The Weibull's cumulative distribution function (CDF) can be expressed in the closed-form [368]

$$F(x) = 1 - e^{-\alpha x^\beta}, \, for \, x \geq 0 \tag{5.3}$$

for $\alpha > 0$ and $\beta > 0$ such that the mean of the probability density function is defined as

$$\mu = \alpha^{-1/\beta} \Gamma\left(1 + \frac{1}{\beta}\right)$$

(5.4)

and the variance is

$$\sigma^2 = \alpha^{-2/\beta}\left\{\Gamma\left(1 + \frac{2}{\beta}\right) - \left[\Gamma\left(1 + \frac{1}{\beta}\right)\right]^2\right\}$$

(5.5)

Stress-controlled statistical fatigue lifetime model

For describing a stress-dependent relationship, such as in the case of fatigue strength in the S-N curves, a three-parameter Weibull distribution is utilized further in the form of

$$F(x; \lambda, \delta, \beta) = 1 - exp\left[-\left(\frac{x - \lambda}{\delta}\right)^\beta\right]$$

(5.6)

where $x > \lambda$, $-\infty < \lambda < \infty$ (the location parameter), $\delta = \alpha^{-1/\beta} > 0$ (the scale parameter), $\beta > 0$ (the shape parameter). Therefore, the mean and variance reduce to

$$\mu = \lambda + \delta\Gamma\left(1 + \frac{1}{\beta}\right)$$
$$\sigma^2 = \delta^2\left\{\Gamma\left(1 + \frac{2}{\beta}\right) - \left[\Gamma\left(1 + \frac{1}{\beta}\right)\right]^2\right\}$$

(5.7)

and the corresponding percentiles are

$$x_p = \lambda + \delta[-log(1 - p)]^{1/\beta}, 0 \le p \le 1$$

(5.8)

This formulation has the benefit of being stable in terms of location and scale mapping, which helps uniquely differentiate between closely related fatigue strength scenarios. This means that when a Weibull random variable such as applied stress or defect distribution is converted using location and scale transformations, the resultant variable, such as fatigue lifetime, is another Weibull random variable with a new location and scale parameters [372, 377]. The initial stage in fatigue modeling is to reproduce a material's fatigue behavior across a specified stress range and stress intensity. This was achieved through the experiments (section 3.3.4) and the calibrated models on the experimental basis in chapter 4. The outcome represents the training data of the presented machine learning algorithm. I would like to underline the main assumption for this model, which is the fatigue strength random nature; otherwise, each specimen subjected to constant stress would have the same fixed lifetime. This is never the case; therefore, the applied deterministic models can never be entirely accurate. Hence, the random behavior of this fatigue strength has to be mimicked. Important to highlight is the weakest link principle which states that a structure is strong only as strong as its weakest element [378]. Therefore, the applied distributions must be stable at regimes of fatigue strength where high nonlinearity is expected across load, geometry, length, or size. Therefore, the applied machine learning model here will have dynamic bias and weight parameters, unlike the traditional statistical models, which have static parameters. It is implied then that the algorithm will find the most suitable bias and weight factors in a given fatigue regime without being explicitly programmed to do so. That depends on a property that is not limited to the Weibull probability density function (PDF) but extends to another type which is the Gumbel distribution. The minimal Gumbel model is expressed as

$$F(x; \lambda, \delta) = 1 - exp\left[-exp\left(\frac{x - \lambda}{\delta}\right)\right] \qquad (5.9)$$

where $x \in \mathbb{R}$; $-\infty < \lambda < \infty$ (the location parameter), $\delta = \alpha^{-1/\beta}$ (the scale parameter). As far as central tendency and dispersion are concerned, the mean and variance will be [372]

$$\mu = \lambda - 0.57772 \cdot \delta; \ \sigma^2 = \pi^2 \delta^2 / 6 \qquad (5.10)$$

and the percentiles for Gumbel PDF will be

$$x_p = \lambda + \delta[log(-log(1 - p))], 0 \le p \le 1 \qquad (5.11)$$

Satisfaction of compatibility conditions
The stress-life model is derived based on the compatibility condition of Castillo and Cantelli [372]. In a Woehler field, the compatibility of CDF of the lifetime N_f and the stress range $\Delta\sigma$ must be ensured such that the minimization law of the weakest link principle is valid

$$Q^*(N^*, \Delta\sigma^*) = E^*(N^* | \Delta\sigma^*) = F^*(\Delta\sigma^* | N^*) = q_{min}(N^*, \Delta\sigma^*) \qquad (5.12)$$

It is implied then that finding the compatible pairs of life and stress is a minimization optimization problem. The literature shows how the function is solved [374] leading to the log-dimensional Weibull model for which the maximum likelihood must be inversely minimized

$$Q(N, \Delta\sigma) = 1 - exp\left\{-\left[\tfrac{(\log N - B)\,(g(\Delta\sigma) - C) - \lambda}{\delta}\right]^{\beta}\right\};$$

$$(\log N - B)(g(\Delta\sigma) - C) = constant \qquad (5.13)$$

such that B is related to the fatigue lifetime threshold and C is related to the fatigue endurance limit. However, these two biases will be treated in this thesis as dynamic and will be automatically adjusted for every stress based on the fatigue physical principle that there is no fatigue limit [379]. The zero-percentile curve indicates the smallest number of cycles required to reach fatigue failure for various stress levels, which is the value presented by the bias in lifetime or load. In this case, the bias in a lifetime might be viewed as the transition point between the fracture initiation and propagation phases. In the case of Gumbel CDF, the solution will be

$$Q(N, \Delta\sigma) = 1 - exp\left[-exp\left(\frac{(\log N - B)\,(g(\Delta\sigma) - C) - \lambda}{\delta}\right)\right] \qquad (5.14)$$

For a more generalized case that considers different shape parameters for N and $\Delta\sigma$, it follows that the log-dimensional Weibull model is convoluted into

$$Q(N, \Delta\sigma) = 1 - exp\left[-\tfrac{(\log N - B)^{\beta}\,(g(\Delta\sigma) - C)^{\gamma}}{\delta}\right];$$

$$(\log N - B)(g(\Delta\sigma) - C) \geq 0; \qquad (5.15)$$

$$(\log N - B)^{\beta}(g(\Delta\sigma) - C)^{\gamma} = constant$$

and the corresponding Gumbel model

$$Q(N, \Delta\sigma) = 1 - exp\left[-exp\left(\tfrac{(\log N - B)^{\beta}\,(g(\Delta\sigma) - C)^{\gamma}}{\delta}\right)\right];$$

$$logN \geq B, (g(\Delta\sigma) - C) \geq 0 \qquad (5.16)$$

The zero-percentile curve in generalized three-parameter models will degenerate into two asymptotes as proposed originally by Castillo and Cantelli [372] since the biases here are dynamic and optimized per specific load scenario. This indicates that the minimum number of cycles required to reach fatigue failure will not remain constant as stress changes, implying that failure does not depend only on the applied stress but on further influences such as the distribution of inhomogeneities. That concept is applied in contrast to earlier views in the literature [380]. However, the advantage is that the algorithm gains more flexibility and more generality in its application to all fatigue regimes without a constraint of a fatigue limit. Therefore, there is no necessity that a very tiny percentile should serve as the zero probability of failure. Thus, the adopted concept comes in accordance with the idea that a fatigue limit for engineering metallic materials does not exist [267, 350,379]. According to some researchers, there is no "a priori" reason to exclude model forms that are fulfilling

$$f(N; \Delta\sigma) = 1 - exp\{-LG(N; \Delta\sigma)\},   \qquad (5.17)$$

where G is a suitable rising function of lifetime and stress $\Delta\sigma$, and L is the specimen length or volume [381]. The adopted model here is a handy and practical model that is really not only meeting the weakest link principle prerequisites but also align in accordance with the latest concepts in fatigue science that are emerging because of the development of new testing machines, which make it possible to test beyond the gigacycle lifetimes.

5.1.2   Parameter Estimation

There are several methods and procedures for the estimation of the Weibull and Gumbel parameters. For instance, the maximum likelihood estimation (MLE) is widely used [382]. Another method suggested being used by Hosking for estimating the parameters is the probability-weighted moments (PWM) [383]. A third method was proposed, which was based on the 2-stage procedure of estimating the parameters of Weibull distribution [377]. There are some comparisons between the three proposed methods. For example, the PWM was found to be outperforming the MLE in several cases, but this applies only to cases where the shape parameter $\beta < 2$ [383].

Maximum likelihood estimation (MLE)
Maximum likelihood estimate is a critical technique in all statistical reasoning and parameter estimation in Bayesian statistics. I shall not go into a detailed description of the approach since there is a lot of literature on widely different applications

handling this topic. Rather than that, I will attempt to explain the maximization concept and demonstrate how it can be applied to the S-N field problem. As the name indicates, the maximum likelihood approach is one that maximizes the likelihood function. The probability function is best described using a discrete distribution and a single parameter as an example [368, 384]. The independent random variables N_1, N_2, ..., N_n are drawn from a discrete probability distribution denoted by $f(N, \Delta\sigma)$, where $\Delta\sigma$ is a single parameter of the distribution. Now the likelihood is described as

$$L(N_1, N_2, \ldots, N_n; \Delta\sigma) = f(N_1, N_2, \ldots, N_n; \Delta\sigma)$$
$$= f(N_1, \Delta\sigma) f(N_2, \Delta\sigma) \cdots f(N_n, \Delta\sigma) \qquad (5.18)$$

which is the joint probability product of the random variables' distributions associated with fatigue-related influences, which is frequently referred to as the likelihood function. Take note that the probability function's variable is $\Delta\sigma$, not N. The observed values in a sample are denoted by N_1, N_2, ..., N_n. The meaning is straightforward in the case of a discrete random variable. The likelihood of the sample, denoted by the value of the function $L(N_1, N_2, ..., N_n; \Delta\sigma)$ is equal to the following joint probability

$$P(N_1 = N_{f1}, N_2 = N_{f2}, \ldots N_n = N_n | \Delta\sigma) \qquad (5.19)$$

In other words, it represents the chances of receiving the sample values N_1, N_2, ..., N_n given a $\Delta\sigma$. In the discrete situation, the maximum likelihood estimator is one that maximizes the joint probability or the sample's chance. Given independent observations N_1, N_2, ..., N_n from a probability density function (continuous case) or probability mass function (discrete case) $f(N; \Delta\sigma)$, the maximum likelihood estimator $\widehat{\Delta\sigma}$ is that which maximizes the likelihood function [370]

$$L(N_1, N_2, \ldots, N_n; \Delta\sigma) = f(N; \Delta\sigma)$$
$$= f(N_1, \Delta\sigma) f(N_2, \Delta\sigma) \ldots (N_2, \Delta\sigma) \qquad (5.20)$$

The continuous probability distribution (CDF) of fatigue lifetime is denoted by $f(N, \Delta\sigma)$ Weibull formulation would be [372]

$$f(N, \Delta\sigma) = 1 - exp\left[-\left(\frac{N - \lambda}{\delta}\right)^{\beta}\right]; N \geq \lambda \qquad (5.21)$$

where λ, δ and β are the corresponding Weibull law parameters. Hence, the reverse Gumbel distribution is the limit of Weibull distribution, Gumbel is formulated as

$$f(N, \Delta\sigma) = 1 - exp\left[-exp\left(\frac{N - \lambda}{\delta}\right)\right]; N \in \mathbb{R} \qquad (5.22)$$

The maximum likelihood approach enables the analyst to make use of prior information about the distribution of fatigue strength in order to select a suitable estimator. As the number of fatigue lifetime observations increases, the maximum likelihood estimator becomes unbiased asymptotically or in the limit; that is, the number of biases approaches zero, and the relation between an expected lifetime and applied loading conditions becomes more precise. A detailed study of the characteristics of maximum likelihood estimation is typically a primary focus of a course on statistical inference theory. However, quite frequently, it is more convenient to work with the log-likelihood of the probability function when determining the parameters that maximize the precision in the relation between stress and lifetime. For a sample $(N_1, N_2, ..., N_n)$, the maximum likelihood estimates of the parameters of the Weibull model are deduced by maximization of the log-likelihood would be [368, 372]

$$L = - \sum_{i \in I1 \cup I0} \left(\frac{N_i - \lambda}{\delta}\right)^{\beta} + (\beta - 1) \sum_{i \in I1} \left(\frac{N_i - \lambda}{\delta}\right) + \sum_{i \in I1} log\frac{\beta}{\delta} \qquad (5.23)$$

with respect to λ, δ and β, where I_0 and I_1 are the set of runouts and non-runouts, respectively. Following the same concept, the Gumbel likelihood will be [368, 372]

$$L = - \sum_{i \in I1 \cup I0} exp\left(\frac{N_i - \lambda}{\delta}\right) + \sum_{i \in I1} \left(\frac{N_i - \lambda}{\delta}\right) + \sum_{i \in I1} log\delta \qquad (5.24)$$

Maximization of the expressions Equation 5.23 and Equation 5.24 would yield estimators of λ, δ and β for Weibull and Gumbel models, respectively, hence the compatibility condition can be secured.

Estimation of the threshold values

The threshold values B and C are related to the threshold fatigue lifetime N and endurance limit of $\Delta\sigma$. As expressed earlier, these are the biases of the algorithm, which will be in this scheme dynamic biases based on the requested stress amplitude. The mean value of $f(N, \Delta\sigma)$ is the regression curve of N on $\Delta\sigma$, which will be deduced based on the FEM scheme of chapter 4, therefore [372, 377]

$$E[logN - B|g(\Delta\sigma) - C] = \frac{\mu}{g(\Delta\sigma) - C} \qquad (5.25)$$

such that

$$E[logN|g(\Delta\sigma) - C] = B + \frac{\mu}{g(\Delta\sigma) - C} \qquad (5.26)$$

Minimization of the regression equation with respect to the applied stress amplitude leads to the dynamic biases B and C which in turn lead to Q

$$Q = \sum_{i=1}^{n} \left(logN_i - B - \frac{\mu}{g(\Delta\sigma_i) - C} \right)^2 \qquad (5.27)$$

where n is the sample size and N_i is the number of cycles to failure of the ith specimen tested at stress range $\Delta\sigma_i$.

5.1.3   Application of the Scheme

To implement sections 1.1.1 and 1.1.2, a MATLAB routine was written, which uses the input data of chapter 4 as predicted trends of fatigue results and initializes stress ranges of interest and corresponding fatigue lifetimes. As an absolute minimum, to start this analysis, three different stress range values with their corresponding fatigue lifetimes must be initialized from the modeling results. At this stage the position is suitable to apply the MLE according to Equation 5.23 and Equation 5.24, using optimset('MaxFunEvals',10000, 'MaxIter', 10000). It returns options that have been configured with the provided parameters through one or more name-value pair arguments. fminsearch begins at x_0 and seeks the local minimum x_i of the function specified. As a result, using the optimization choices supplied in the general-purpose structural options of MATLAB, minimization of the bias of the algorithm can be achieved. To configure Weibull parameters, optimset is used for the MLE. With this output at hand, the elements to establish the Weibull model in Equation 5.13 to Equation 5.16 are available. A comparison between experimental and calculated percentiles can be seen in Figure 5.2. In both cases, a good agreement was found between calculated limits

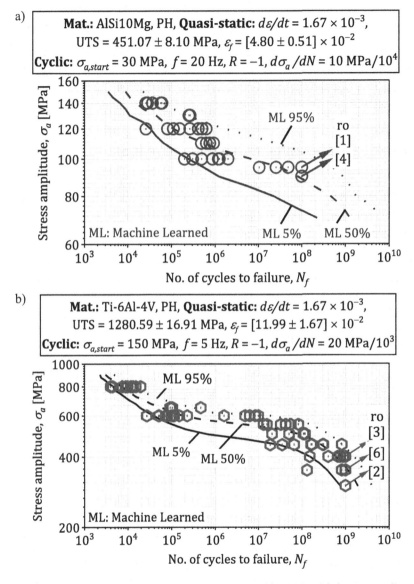

Figure 5.2 Comparison between experimental data and the calculated fatigue percentiles: a) AlSi10Mg; b) Ti-6Al-4V

by the machine learning algorithm and the extremities of the experimental values. This was found consistent in all fatigue regimes from LCF into VHCF, which is a testimony to the success of the concept of dynamic biases and the non-explicit optimization of biases through the proposed machine learning algorithm. This redeems a significant shortcoming of the approach of Castillo and Cantelli [372], where they reported that their model would be suitable only for the HCF fatigue regime only.

5.2   A Two-parameter Bayesian Inference

The weakness of the joint probability formulation is that it tends to maximize the likelihood, which results in an overshoot of the variance that affects the calculation of the percentiles of $Q(N, \Delta\sigma | M, P)$ where M and P are the corresponding probability distributions of microstructure and porosity respectively. Hence, an overshoot of the expected fatigue strength will happen in the ML algorithm. Moreover, the central tendency of expected fatigue strength becomes biased towards local maxima, which is not necessarily physically true and reflects a correct fatigue strength. The origin of this setback lies within the nature of joint probability density functions since they are not subjective density functions of the factors that influence fatigue strength. As a matter of fact, in this thesis, the author tries to boost efficiency by using a limited number of fatigue experiments to none; therefore, the inference of fatigue lifetime of a sample must depend on subjective bias of a fatigue-related influence such as porosity and microstructure. Hence, the conditional perspective that is based on the actual FE predicted values based on experimental fatigue results is given a higher weight since it complements the whole computational scheme.

5.2.1   Statistical Parameter Mapping

Formulation
Frequently in fatigue, the probability distribution of fatigue lifetime comes to be dependent on fatigue-related influences such as surface finish, remnant defects, and more directly, stress concentrations and intensity factors. Let us consider the following example: $\Delta\sigma$ is a parameter of the continuous probability distribution of fatigue lifetime $f(N, \Delta\sigma)$ and the lifetime $N = u(\Delta\sigma)$ specifies a point-to-point mapping between the values of N and $\Delta\sigma$. The interest is in determining the probability distribution of N. It is critical to understand that the point-to-point

transformation means that each value $\Delta\sigma$ is connected to exactly one value $N = u(\Delta\sigma)$ and that each value N is related to exactly one value $\Delta\sigma = w(N)$, where $w(N)$ is derived by solving $N = u(\Delta\sigma)$ for $\Delta\sigma$ in terms of N. The concept can then be described by the relationship [368, 370]

$$g(N) = P(N = N) = P[\Delta\sigma = w(N)] = f[w(N)] \tag{5.28}$$

In case of several factors with an influence on the fatigue lifetime N. is included in an analysis, the rule then must be extended to accommodate a joint probability formulation in the case factors, such as microstructure M, are included. Therefore, $g(N_1, N_2)$ is related to $u_1(\Delta\sigma_1, \Delta\sigma_2)$ and $u_2(M_1, M_2)$ with a unique transformation between related conditions such that the joint probability density function is [368, 371]

$$g(N_1, N_2) = f[u_1(\Delta\sigma_1, \Delta\sigma_2), u_2(M_1, M_2)] \tag{5.29}$$

The function can be solved with a unique Jacobian mapping

$$\begin{aligned} g(N) &= f[w(N)]|J| \\ g(N) &= \sum_{i=k}^{k} f[w_i(N)]|J_i| \end{aligned} \tag{5.30}$$

where $J_i = w_i(N), i = 1, 2, ..., k$, and $J = w'(N)$ and is called the Jacobian of the mapping in case one fatigue-influencing factor has to be mapped on the fatigue lifetime. In case of further factors included, a general case scenario is handled by the determinant of the Jacobian, which is in case of two factors will be [368, 385]

$$J = \begin{vmatrix} \frac{\partial \Delta\sigma}{\partial N_1} & \frac{\partial \Delta\sigma}{\partial N_2} \\ \frac{\partial M}{\partial N_1} & \frac{\partial M}{\partial N_2} \end{vmatrix} \tag{5.31}$$

$\partial\Delta\sigma \big/ \partial N_1$ is simply the derivative of $\Delta\sigma_1 = w_1(N_1, N_2)$ with respect to N_1 with N_2 held constant, referred to in calculus as the partial derivative of $\Delta\sigma_1$ with respect to N_1. The remaining partial derivatives are defined similarly. The transformation of variables is necessary for the application at hand in this work since the distribution of stress, porosity, and microstructural features are available in enormous quantities from FE simulations in chapter 4 and μ-CT and microscopy from section 3.3. The amount of data is enough to produce well-representative density functions. It is the fatigue data that is always at a premium to express enough precision when comparing similar material configurations.

5.2.2   Joint Probability of Fatigue-related Influences

Formulation
In section 1.1, fatigue lifetime distribution was handled as one-dimensional sample spaces, in which the experiment or numerical results are associated with values assumed by a single dependent parameter mapped from one or more fatigue-related influences. However, the more realistic case is the dependence of the distribution of fatigue lifetimes on many influences in a single case at a time. For instance, we are interested here in fatigue lifetimes N_f when a stress range $\Delta\sigma$ is applied during a stress-controlled fatigue experiment of a specimen of Ti-6Al-4V with internal porosity and fine acicular martensite α', resulting in a distribution of fatigue lifetimes $Q(N, \Delta\sigma)$. If we have here two distinctive random variables of fatigue-related influences such as microstructure M and porosity P presented by two probability density functions. We assume that the two factors are influential on $Q(N, \Delta\sigma)$ at the same time with different weights. Then a discrete case of joint probability between them will be [368, 371]

$$f(N, \Delta\sigma|M, P) = P(\boldsymbol{M} = M, \boldsymbol{P} = P) \tag{5.32}$$

The probability mass function of the microstructure M and the porosity P as fatigue-related influences for any region in the S-N field would be

$$\boldsymbol{Q}(N, \Delta\sigma|M, P) = \sum_N \sum_{\Delta\sigma} f(N, \Delta\sigma|M, P)$$

$\rightarrow$ in a discrete case

$$\boldsymbol{Q}(N, \Delta\sigma|M, P) = \int_N \int_{\Delta\sigma} f(N, \Delta\sigma|M, P) \, dM \, dP \tag{5.33}$$

$\rightarrow$ in a continuous case

Although this formulation is a further step towards generalization of the implementation of the relation $Q(N, \Delta\sigma)$, its algorithmic treatment is rather complicated and demanding. Therefore, the formulation can be simplified if statistical independence can be established between the fatigue-related influences. Hence, the author advises considering the case of marginal distributions of M and P alone such that [368, 385]

$$g(M) = \sum_y f(M, P) \, and \, h(P) = \sum_x f(M, P)$$

$\rightarrow$ in a discrete case

$$g(M) = \int\limits_{-\infty}^{\infty} f(M, P) dx \, and \, h(P) = \int\limits_{-\infty}^{\infty} f(M, P) \, dy \qquad (5.34)$$

$\rightarrow$ in a continuous case

Now, we must define fatigue-related influences as functions of each other. For instance, the probability density function of the fatigue-related porosity when microstructure is fixed $f(M, P)/g(M)$. This would also be valid since $f(M, P)$ and $g(M)$ are the joint density and marginal distributions, respectively. This is the same formulation as a conditional distribution. Thus, the conditional distribution of the porosity P when microstructure M is known would be [368, 371]

$$f(P|M) = \frac{f(M, P)}{g(M)}, \, under \, condition \, that \, g(M) > 0 \qquad (5.35)$$

and vice-versa the conditional distribution of the microstructure M when porosity P is known would be

$$f(M|P) = \frac{f(M, P)}{h(P)}, \, under \, condition \, that \, h(P) > 0 \qquad (5.36)$$

If $f(M|P)$ does not depend on P, it follows that $f(M|P) = g(M)$ and $f(M, P) = g(M)h(P)$ which satisfies the condition of statistical independence and provide a more practical formulation for all range of (M, P)

$$Q(N, \Delta\sigma|M, P) = g(M)h(P) \qquad (5.37)$$

Numerical handling

The data treatment begins by declaring the variables of the fatigue-related influences in the form of gamma-based probability distributions with their own scale, shape, and location parameters. The joint expression of Equation 5.37 is then maximized to obtain the joint probability function $f(N, \Delta\sigma | M, P)$ of the regression $Q(N, \Delta\sigma | M, P)$.

5.2.3   Belief Function of Fatigue Strength

Formulation

Let $\mathbb{M}$ be a sample of microstructure M such that $\mathbb{M} = \{M_1, M_2, \ldots, M_k\}$, the following must apply for this specific sample [371, 385]

$$\sum_{k=1}^{K} P(M_k) = 1 \tag{5.38}$$

where the marginal probability is

$$P(M) = \sum_{k=1}^{K} P(M \cap M_k) = \sum_{k=1}^{K} P(M|M_k)P(M_k) \tag{5.39}$$

where the subjective probability according to Baye's rule, will be

$$P(M_j|M) = \frac{P(M|M_j)P(M_j)}{P(M)} = \frac{P(M|M_j)P(M_j)}{\sum_{k=1}^{K} P(M|M_k)P(M_k)} \tag{5.40}$$

The expectation of a certain microstructural state M in the continuous density functional case would be

$$E[M] = \int_{M \in \mathbb{M}} M P(M) dM \tag{5.41}$$

Especially in the gamma distribution types that are being utilized here, the skewness means that the desired output is likely to be far away from the sample mean fatigue lifetime. So, a refrain from utilizing the mean values in any prediction is thoroughly implemented with a focus on the median instead. For the two parameters of interest here, microstructure M and porosity P, there exist two unique and not necessarily statistically independent probability distributions. The marginal density of M can be deduced from the joint density

$$p_M(m) \equiv P(M = m) = \sum_{p \in P} P(\{M = m\} \cap \{P = p\})$$

$$\equiv \sum_{p \in P} p_{MP}(m, p) \qquad (5.42)$$

The conditional density of porosity P resulting from the applied SLM parameters when $\{M = m\}$ is derived from the joint and marginal densities

$$p_{P|M}(p, m) = \frac{P(\{M = m\} \cap \{P = p\})}{P(M = m)} = \frac{p_{MP}(m, p)}{p_M(m)} \qquad (5.43)$$

Now the author might treat $f(N, \Delta\sigma)$ as continuous and M as discrete. Bayesian estimation of $f(N, \Delta\sigma)$ derives from the calculation of $p(N, \Delta\sigma|m)$, where m is the observed value of M in a given SLM printed sample of specimens. This calculation initially requires that we have a joint density $p(N, \Delta\sigma, M)$ representing our beliefs about $f(N, \Delta\sigma)$ and the study outcome of the microstructure M. Often it is trivial to construct this joint density from

– $p(N, \Delta\sigma)$, beliefs about $f(N, \Delta\sigma)$.
– $p(m|N, \Delta\sigma)$, beliefs about the microstructure M for each input of $f(N, \Delta\sigma)$.

Having observed a discrete case of the microstructure $\{M = m\}$, we must change our views and expectations regarding $f(N, \Delta\sigma)$, such that [385, 386]

$$p(N, \Delta\sigma|m) = p(N, \Delta\sigma, m)/p(m)$$
$$= p(N, \Delta\sigma)p(m|N, \Delta\sigma)/p(m) \qquad (5.44)$$

This conditional density is called the posterior density of $f(N, \Delta\sigma)$ which we aim to maximize its precision in the algorithm. Suppose $(N, \Delta\sigma)_a$ and $(N, \Delta\sigma)_b$ are two possible combinations of numerical value events in the S-N field of the true value of $f(N, \Delta\sigma)$. The posterior probability (density) of $(N, \Delta\sigma)_a$ with respect to $(N, \Delta\sigma)_b$, conditional on a discrete case of the microstructure $M = m$, is

$$\frac{p((N, \Delta\sigma)_a|m)}{p((N, \Delta\sigma)_b|m)} = \frac{p\big((N, \Delta\sigma)_a\big)p(m|(N, \Delta\sigma)_a)/p(m)}{p\big((N, \Delta\sigma)_b\big)p(m|(N, \Delta\sigma)_b)/p(m)}$$

$$= \frac{p\big((N, \Delta\sigma)_a\big)p(m|(N, \Delta\sigma)_a)}{p\big((N, \Delta\sigma)_b\big)p(m|(N, \Delta\sigma)_b)} \tag{5.45}$$

This means that to evaluate the relative posterior probabilities of $(N, \Delta\sigma)_a$ and $(N, \Delta\sigma)_b$, we do not need to compute the actual PDF of $p(m)$. Another way to look at it is, as a function of $f(N, \Delta\sigma)$

$$p(N, \Delta\sigma|m) \propto p(N, \Delta\sigma)p(m|N, \Delta\sigma) \tag{5.46}$$

The constant of proportionality is $1/p(m)$, which could be calculated from

$$p(m) = \int_M p(m, (N, \Delta\sigma))d(N, \Delta\sigma)$$

$$= \int_M p(m|N, \Delta\sigma)p(N, \Delta\sigma)d(N, \Delta\sigma) \tag{5.47}$$

What will be aimed for in this section will be a posterior probability of fatigue lifetime based on the subjective bias rather the conditional experimental perspective of fatigue lifetime. Hence, based on Equation 5.46 and Equation 5.47, then the percentiles of the distribution of $f(N, \Delta\sigma)$ given the microstructure M or porosity P would be [368, 371]

$$Q(N, \Delta\sigma|M) = \frac{f(M|N, \Delta\sigma)\pi(N, \Delta\sigma)}{g(M)} \tag{5.48}$$

where $\pi(N, \Delta\sigma)$ is the posterior distribution of the stress-life conjugate $f(N, \Delta\sigma)$ and $g(M)$ is the marginal distribution of the microstructure M. Equation 5.48 will be used interchangeably with the porosity P. In this thesis, the focus is on studying the influence of microstructure M and porosity P exclusively and as an interactive effect, which interactively can be expressed as

$$Q(N, \Delta\sigma|M, P) = \frac{f(M|N, \Delta\sigma)f(P|N, \Delta\sigma)\pi(N, \Delta\sigma)}{g(M)h(P)} \tag{5.49}$$

where $h(P)$ is the marginal distribution of the porosity P.

5.2.4 Estimation of Error

In Bayesian analysis, the posterior distribution of a fatigue-related influence can be calculated. When a loss of accuracy occurs, the posterior distribution and a loss function can also be used to construct Bayes estimates. A loss function measures the costs of making a certain statistical decision compared to the target fatigue lifetime. Thus, error minimization would be the optimization function in the ML algorithm

$$L\left(N_f^a(M), N_f\right) = \left(N(M) - N_f\right)^2 \tag{5.50}$$

where M is the microstructure, N_f^a is the expected fatigue lifetime and N_f is the fatigue lifetime of the training data. The goal of the Bayesian inference analysis is to minimize the error loss function for the fatigue lifetime. The mean of the posterior distribution of microstructure given a fatigue-related lifetime $\pi(N_f^a(M)|N_f^a)$, denoted by $N_f^a(M)^*$, is the Bayes estimate of microstructure M influenced fatigue lifetime $N_f^a(M)$ under the squared-error loss function of the training fatigue lifetime. Furthermore, a linear deviation from training fatigue lifetime values can also be given by the absolute error.

$$L\left(N_f^a(M), N_f\right) = \left|N_f^a(M) - N_f\right| \tag{5.51}$$

where M is the microstructure, N_f^a is the expected fatigue lifetime and N_f is the fatigue lifetime of the training data. The median of the posterior distribution of microstructure given a fatigue-related lifetime $\pi(N_f^a(M)|N_f)$ denoted by $N_f^a(M)^*$, which is the Bayes estimate of microstructure M influenced fatigue lifetime $N_f^a(M)$ under the squared-error loss function of the training fatigue lifetime.

5.3 Influence of Structure and Process on Fatigue Strength

5.3.1 Remnant Defects

Using the machine learning (ML) algorithm based on Bayesian statistics in sections 1.1 and 1.2, the defect-correlated assessment of fatigue strength was established. The defect distributions of the different specimens of the batches of

Table 3.2 were used as an input to the ML program. In Figure 5.3, a contour plot of the related fatigue strength across the whole range of resulting defects is presented from 10^3 cycles until 10^{10} cycles for a 50% probability of failure. To generalize the resulting strength value, it is represented not in terms of load but in terms of load density per unit volume. This excludes away the influence of specimen geometry and size effect. The reduction in fatigue strength due to an increase in mean defect size is consistent across all studied fatigue decades. However, the reduction across decade 10^4 is steeper. A further advantage is highlighted here is that there is an exclusion of the fatigue limit concept whereby the fatigue strength was possible to map into several Giga cycles of fatigue strength, unlike earlier shortcomings [372].

In comparison to the literature, we see an emphasis on studying the influence of loading conditions on fatigue strength by ML algorithms. Mean stress is a key element in fatigue design, especially in high cycle service applications. The issue arises from the Fourier transform combining all cycle by cycle mean stress effects into a single zero frequency content. Nonlinear generalization is possible in artificial neural networks [387]. A neuro-fuzzy-based machine learning approach was used to estimate the high cycle fatigue life of laser powder bed fusion stainless steel 316L by entering parameters for processing/post-processing and static tensile properties. However, due to the heterogeneity of the provided data, straight application of the model to it yielded a range of accurate predictions [388]. Total fretting fatigue life is predicted using contact size, peak pressure, remote specimen tension, and tangential force ratio. 90% of the data is used to train, and testing of the artificial neural network follows. The network is shown to be extremely good at discriminating between low-life and 'run-out' outcomes. It does less well in predicting the difference between low and high life [389]. Uncertainties arising from geometry, material, and models are extensively quantified using data from measurements and tests to estimate the life of fatigue crack growth (FCG) on turbine discs. Uncertainty is quantified using a Bayesian method. The Gaussian process regression technique was introduced to describe the accuracy and computational efficiency of high model accuracy measurements [390].

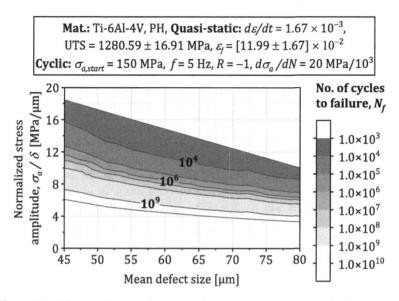

Figure 5.3   Influence of mean defect size on fatigue strength as computed by the proposed machine learning algorithm at 50% probability of failure

For the estimate of probabilistic fatigue *S-N* curves, a probabilistic physics-guided neural network (PPNN) was suggested. The suggested model addressed limitations inherent in existing parametric regression and traditional ML techniques for the interpretation of fatigue data. In comparison to explicit regression-type models, the PPNN is more adaptable and does not impose constraints on function types under various stress levels, mean stresses, or other variables [391]. The fatigue damage behavior of AM aerospace alloys was investigated using a novel technique. The continuum damage mechanics (CDM) theory was efficiently integrated with ML models [392]. A multi-scale FEA and ML framework for predicting fatigue performance in welded structures was suggested. Residual stress and toe radius were used to integrate past information about microstructures and characteristics, as well as uncertainties of fatigue life [393].

5.3.2   Width of Prior β Grains

Traditionally, the influence of characteristics of grain sizes on fatigue strength has been quantified in literature through numerically demanding procedures such as CPFEM [394]. On the other hand, artificial intelligence (AI) methods offer a numerically efficient approach for mapping fatigue properties to microstructural features. In Figure 5.4, the mapping of the size of the prior β grain to the fatigue strength from 10^3 to 10^{10} cycles for the load density applied per unit volume is given. The computation was based on the probability distributions of the sizes of the prior β grain sizes resulting from in situ treatments through variation of second exposure treatments in Table 3.2. It was observed that the fatigue strength dependence on grain size is steep until around 120 μm, after which the slope of the dependence flattens and stabilizes. The map produced through the presented ML algorithm is across the full range and into tens of giga cycles of fatigue.

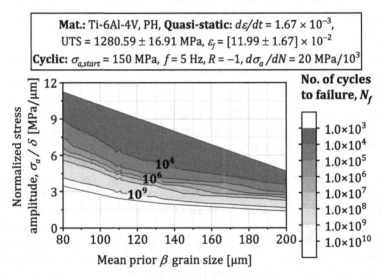

Figure 5.4   Influence of mean prior β grain size on fatigue strength as computed by the proposed machine learning algorithm at 50% probability of failure

As mentioned above, micromechanical models accurately represent material behavior and its relationship to microstructural processes; however, they are numerically expensive. Although such models can be used to train the ML models, when out-of-range data were used to train machine learning models, projected grain sizes varied significantly from reference values. Supervised learning had a weaker capacity to generalize to previously unknown data than random forest regression (RFR) when it came to damage evolution prediction [395]. A framework was created for data-driven analysis of AM stainless steel 316L fatigue life prediction using continuum damage mechanics (CDM) to characterize the impacts of AM process parameters [396]. Logistic regression ML was also used to recognize fatigue failure in cyclically damaged structures. The results show that failure can be identified before a technical fracture appears and that maintenance interval could be optimized [397]. A deep learning framework is used to build a neural network (NN), which was trained and tested to make predictions of fatigue strength for the simulated data of input parameters. This was then used to generate conditional probability tables for the Bayesian network (BN) [398].

5.3.3 Energy Density Ev

The experimental and numerical analysis of fatigue strength in this thesis focused on the quantification of process-related defects and microstructural features. It is not debatable that both are a direct descendent of laser process parameters and, in particular, the energy density [7, 231]. The Bayesian algorithm in section 1.2 was used to map the fatigue strength of various applied energy densities from Table 3.2. Figure 5.5 shows a map between energy density and load density per unit volume for the various fatigue regimes from 10^3 cycles until 10^{10} cycles. The map focuses on the range of applied energy densities that were applied in the current study. Further extrapolations into virtual predictive scenarios can be based on the training data when predicted values have a sufficient physical and statistical relation to the training data. The large energy input necessitates a high cooling rate during processing which might not always be beneficial for mechanical properties. Novel techniques have been developed that attempt to maintain a high degree of energy input while controlling cooling rates in order to tune the microstructure and decrease residual stresses [399]. Cumulative energy absorption increases as the power of energy function increases [400]. Rapid cooling results in the formation of an intermetallic Ti3Al phase which are more stable in high-temperature application but not desirable at room temperature [401]. If the lower energy density is used, higher residual stress (RS) is generated in the subsurface region [402]. Parameter sets with an excessively high or insufficiently low energy input are unfavorable for mechanical properties. In SLM, the higher energy density drives the Ti-6Al-4V powder bed to over-melt [403]. It is planned in future studies following this thesis to include further effects such as residual stress mapping on fatigue strength. Further extensions of the presented model would attempt to forecast fracture behavior and expected crack growth rates of certain microstructures resulting from several process parameter sets. Some efforts in this direction are already showing up in the literature. A trained network could forecast the remaining of a specimen's crack development based on the initial half of the data points. The technique outperforms traditional fitting models [404]. An ANN-based model was created to assess high-cycle fatigue fracture development rates (da/dN) for dual-phase (DP) steel. The ANN-based model simplifies the complexity of test data and parameter relationships [405]. Radial basis function neural network (RBF-NN) was used to tackle the critical fatigue crack propagation in aircraft structures that have a degree of nonlinearity even in the Paris region. On the other hand, RBF-NN was much more flexible to handle nonlinearity [406].

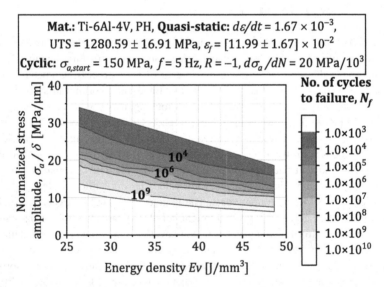

Figure 5.5 Influence of total applied energy on fatigue strength as computed by the proposed machine learning algorithm at 50% probability of failure

Summary and Outlook

6

6.1   Processing

Recognizing the mechanisms that cause leftover porosity to form during selective laser melting aids in the development of techniques to eliminate the porosity that is harmful to dynamic structural applications. Following a review of the literature, it was determined that melt pool instabilities and high cooling rates cause turbulent melt pools to collapse over pressure-decreasing voids. To this aim, the author recommended using secondary exposure of the present build layers to improve degassing before solidification and allow for appropriate and stable heat accumulation over the specimen build-up, particularly in the vertical direction. The work used two common lightweight alloys, AlSi10Mg and Ti-6Al-4V, from the selective laser melting material palette. Laser power, scanning speed, spot size, and energy density were used to create a remelting designed experiment to variate porosity and microstructure. The material bed is exposed to a second exposure in accordance with the specified experiment after an optimal standard initial exposure on the construction layer.

A thermal camera tracked the cross-sections of the specimens during the vertical build-up, measuring the mean surface temperature in the region of interest. Before recoating with powder and rescanning with the laser, the measurements were recorded in two steps. Heat accumulation was substantial in most remelting parameter sets when compared to single exposure mean surface temperature, which could also reach much greater values over the build layer and sustain the achieved temperature for extended periods of time. The mean surface temperature of the mechanical testing specimens, which were 10 times longer in height, shows a sharp positive gradient up to 20 mm, after which the gradient decreases and

© The Author(s), under exclusive license to Springer Fachmedien Wiesbaden GmbH, part of Springer Nature 2022
M. Mamduh Mustafa Awd, *Machine Learning Algorithm for Fatigue Fields in Additive Manufacturing*, Werkstofftechnische Berichte | Reports of Materials Science and Engineering, https:doi.org/10.1007/978-3-658-40237-2_6

the temperature stabilizes throughout construct layers. The temperature profile throughout a region of interest was unique, having two temperature maxima for the first and second exposures. The second exposure helped maintain the thermal energy and reduced the cooling rate, allowing the melt pool to be exposed to high temperatures for a longer period. X-ray microcomputed tomography was used to assess the efficacy of the remelting technique, providing detailed information on residual porosity in a variety of forms. The applied remelting technique was successful in decreasing both the number of pores and the overall defect volume for Ti-6Al-4V in most of the cases. In comparison to the other parameters, the energy density parameter was the most effective.

However, for functionally graded microstructure by remelting, it is critical to understand the relationship between different scanning vector configurations and the resultant defect distribution. As a result, the transition zones between processing regions in graded systems are of special importance. To understand their impact on porosity development in the transition zone between two processing conditions, specimens with three different vector overlaps/gaps in the transition zone should be created. Additionally, to print samples with continuously graded vectors, a directly programmable scanner system should be utilized, as the inertia of the scanner mirrors causes acceleration and slowdown of the laser beam during transition zones. The influence of the local cooling situation can be examined with a thermal camera at high magnification.

6.2   Mechanical and Structural Properties

X-ray microcomputed tomography (μ-CT) was used to study the effect of process variables on residual porosity. Pore identification was accomplished by viewing the reconstructed volumes using the application VGStudio Max 2.2 (Volume Graphics) and classifying them as defects. The number, position, and size of pores can vary substantially depending on the materials used. Pore formation is a result of Marangoni convection and the torque that it generates. The intensity of the torque is related to the magnitude of the temperature differences in the melt pool area. AlSi10Mg had a relative density of 99.94%. It is clear from all the specimens that sphericity improves as the corresponding energy density increases. At 200 μm, the pore size reaches a maximum. There is a critical energy density point for this alloy, roughly 60 J/mm^3, at which the lowest pore fraction is obtained. Keyhole pores were formed as the scanning speed increased, accompanied by a decrease in metallurgical pores. The observation that the biggest flaws in the SLM material were extremely anisotropic, being flat and disc-like in the xy-plane, is significant since the planes of these pores match the layer-wise manufacturing process planes. A greater fraction of pores exhibited sphericity smaller than 0.5 in Ti-6Al-4V than in AlSi10Mg. The extra heat from the platform does not have the same impact as it does in AlSi10Mg because the applied 200 °C is insufficient for Ti-6Al-4V. According to Awd et al. [183], the proposed remelting procedure was a successful strategy for decreasing porosity since nearly completely dense specimens could be created.

In this study, AlSi10Mg and Ti-6Al-4V specimens were cut using the Labotom-3 cutting machine. The specimens are ground and polished to prepare the surface for further etching. Metallographic sections were used to depict the phases seen in microstructures observed by light or scanning electron microscopy. Randomly, small-scale porosity on the order of 40 μm in diameter was discovered. Pores observed in the non-platform heated 2D micrographs were similarly dispersed in their position. AlSi10Mg alloy contains nanocrystalline dendritic silicon particles that contribute to its exceptional hardness, strength, and impact toughness. When oriented in a favorable direction, the alloy obtained similar strength and ductility to alloys generated using SLM. SEM was used to magnify the melt tracks. It was found that the development of small-scale porosity in melt pools is distinct from the balling and Marangoni force process. In both alloys, the melt tracks were characterized by the presence of several parallel heat-affected zones at some locations. These spots represent regions of microstructure transition, which results in a distinctive hue when the specimen is seen in light microscopy. The bulk microstructure of Ti-6Al-4V specimens is characterized by columnar prior β grains aligned in the build direction and continuous colonies of Widmanstaetten structure. A favored orientation connection results in the development of 40–60 μm wide lamellar colonies. The decrease in the average lath width at the junction can be attributable to the increased cooling rate during each subsequent layer deposition. In general, the SLM process involves the heating and cooling of the material on a cyclical basis. Heat treatment in situ occurs when additional laser scans are performed on the solidified material after it has been heated to a certain temperature by the lasers. Twining is said to arise because of the high temperatures experienced during manufacturing. The remelting exposure degrades the prior β grain morphology to the extent that they become indistinguishable under light or scanning electron microscopy.

AlSi10Mg had a tensile strength of 451.07 MPa, which is higher than the cast counterpart, whereas Ti-6Al-4V with the employed parameters has a greater tensile strength of 1280.59 MPa and a higher yield strength than wire + arc additively manufactured counterpart. As a result of strain localization, micro-voids develop at the melt tracks and combine to nucleate fractures that propagate across the interface, resulting in mode-I fracture without ductility. Platform heating had no effect on the elastic behavior, according to Awd et al. [191]. Microstructure results of SLM-processed Al-Si alloys with extra-high silicon content showed that the primary silicon phase has a great influence on the flow properties and eliminates hot cracks. Si content and its strengthening Mg phases are the primary reason for the high microhardness of AlSi10Mg. A high-density SLM Ti-6Al-4V had a micro-hardness level of 413 HV, which is higher than the EBM counterpart. Using the ultra-micro instrumented indentation test, the elastic modulus of the microstructure was determined. The dependency of hardness and elastic properties on the true contact area was discussed. The link between the contact depth and the indentation depth was not straightforward since the eventual accumulation or sinking that occurs during an indenter process cannot be observed only by monitoring tip displacement. The value of elastic modulus obtained through instrumented indentation was higher than that obtained by the standard tensile test.

Cyclic deformation under incremental loading for AlSi10Mg starts at a stress amplitude of 30 MPa, and the stress increases at a rate of 10 MPa/10^4 cycles. After reaching 100 MPa, the specimen starts to harden towards fracture, as evident by the slight gradual decrease of plastic strain amplitude. The alloying system of AlSi10Mg develops cyclic hardening in comparison to the $\alpha + \beta$ nature of Ti-6Al-4V. Higher than 450 MPa of stress amplitude, plastic strain amplitude drops slightly and gains a parabolic relationship with loading, which indicates the engagement of further interactions of cyclic deformation mechanisms such as severe strain localizations, microstructural hardening, and formation of a multitude of cracks. At 100 and 120 MPa for AlSi10Mg, the specimens show a continuous decline of elastic and plastic strains, including the amplitude and mean components. The results suggest a change in the fatigue damage mechanism at higher loads when compared to 140 MPa. The amount of total strain is significantly decreased as the test progresses in the LIT of Ti-6Al-4V. This is ought to the higher complexity of the $\alpha + \beta$ microstructure of Ti-6Al-4V. Fatigue failures arose from the surface and sub-surface flaws. Clusters of pores caused fatigue failures in most situations in AlSi10Mg which has been shown to experience cyclic hardening. Porosity plays a critical role in the definition of fatigue strength characteristics of Ti-6Al-4V. Failure was observed at 800 MPa of stress amplitude from a pore that is~250 μm deep from the surface of the specimen. Since ultrasonic vibration in metallic materials generates temperature because of the internal resistance to cyclic deformation, a pulse:pause scheme, in addition to the application of compressed dry air, is applied to the fatigue specimen to control the accumulation of temperature.

The ultrasonic fatigue data of AlSi10Mg was presented from the late HCF regime into the VHCF regime. It was observed that between 140 and 100 MPa, the life does not increase significantly, and the scatter is relatively less in comparison to fatigue strength between 100 and 90 MPa of stress amplitude. At these stress amplitude levels, the fatigue strength is more horizontally distributed with high scatter, and therefore 4 runouts take place at 90 MPa for a limiting number of cycles of 1E8. For Ti-6Al-4V, the life increases rapidly in the range of 650 to 550 MPa of stress amplitude with a low slope S-N relation. At these stress amplitude levels, the fatigue strength is more horizontally distributed with high scatter. When the strain rate is high, yet the load is low, the motion of dislocations is not reaching the prior β boundaries, and the strain is localized within the colonies. The damage is more rapid at this S-N range, and life increase due to low reduction is not as significant as in earlier higher load experiments. 3D crack morphology of an AlSi10Mg specimen showed crack initiation location at a cluster of defects in the subsurface of the specimen. Since load is very low in the VHCF range, the crack propagation line showed linear elastic characteristics, but towards the end of its life, the crack front starts to plasticize and deflect. This could be due to microstructural discontinuities, which exist in plenty of quantity in the vertical direction in AlSi10Mg but are not so evident in Ti-6Al-4V.

6.3   Fracture Mechanics

The XFEM and contour integral techniques in Abaqus can be used to compute the SIF at the crack tip. Parametric research was carried out to investigate SIF accuracy derived by Abaqus as crack length change. The overall accuracy of the contour integral approach is greater than that of the XFEM method for static SIF analysis. When the fracture length exceeds a threshold value, however, the stress intensity becomes excessive, the mesh density of the contour integral technique fails to fulfill the requirement, and the accuracy suffers significantly. The accuracy of XFEM is unaffected by the stress intensity because it is a mesh-independent technique. XFEM was used to model the crack propagation process with an arbitrary crack route to study the fatigue crack development process of additively manufactured fatigue specimens. XFEM, contour integral, and analytical methods can all be used to calculate ΔK values. C and m can be calculated from the simulation results by fitting the curves between ΔK and da/dN. When compared to the contour integral technique, XFEM and Analytical methods offer greater accuracy in fitting parameter m. Analytical technique parameter C is the closest to the experimental value, followed by XFEM. The inaccuracy of $\log C$

calculated using the contour integral technique is the greatest, more than double that of XFEM. It was inferred that the contour integral technique produces more accurate static SIF values in small stress-intensity zones, whereas XFEM is better for large-stress-intensity regions.

To simulate fatigue fracture propagation, a three-dimensional model was used in two stages; the static general step and the direct cyclic step, which are required for a model without a pre-defined crack. The Quade damage initiation criterion was effective for identifying the initial crack location by computing strain values and determining the weakest point. The failure initiates at the pores because the cross-sectional areas of the planes with extensive pores are the smallest. The fracture would merely begin rather than develop rapidly until later since the maximum energy release rate was significantly lower than the critical energy release rate. Because the maximum energy release rate became greater than the threshold with loading cycles, the first crack produced began to propagate under cyclic stress and followed the fatigue crack propagation curve. The rate of fatigue fracture propagation rises over time and follows an energy release law based on the Paris model.

All simulations were performed on a supposedly SLM-built cylindrical fatigue specimen. X-ray microcomputed tomography (μ-CT) scans produce pure 2D scans that are taken in angular increments. The scans are uploaded into VGStudio Max 2.2, which stitches and joins them together to create a 3D shell representation. To reduce the number of faces and vertices, numerous reductions and simplification techniques must be carried out. MeshLab is used for the second simplification procedure, which employs the quadric edge collapse decimation technique. An STL shell representation must be converted to a solid, and internal porosity must be reduced from the solid representation. Separating the cylindrical shell from interior porosities was accomplished using FreeCAD. Crack propagation rate curves were used to study the influence of loading level and testing frequency on fatigue crack propagation rate. In all shown stress amplitudes, the fatigue crack growth area is proceeding at a very slow rate before 1E3 cycles. At higher loads, an abrupt increase in the crack area takes place at around 3E3 cycles, followed by a stable horizon until around 4E4 cycles. After this, the unstable crack propagation zone proceeds towards ultimate failure giving the final number of cycles to failure. Crack propagation rate curves can be used to conclude about Paris constants of this studied specimen geometry in dependence on the applied load. For both alloys, good agreement was found between experiment and simulated outcomes.

6.4   Cyclic Plasticity

The proposed technique allows the identification of isotropic and kinematic hardening parameters on its own without interaction since two independent procedures are used. The fatigue lifetime prediction model based on Fatemi-Socie damage depends on the maximum shear strain developed under simulated cyclic loading. A prior trial in Abaqus had suggested that the flow curve is not enough for simulating the system since the cyclic hardening effect and the Bauschinger effect are not captured. With the support of numerical experiments, only tensile and stress-controlled increasing load tests were included in the parameter optimization. A basic approach is to compare the initial stress amplitude to the newly stabilized stress amplitude. For every material, a wide range of strain levels is applied with larger increments, and a restricted range is formed with smaller increments. The use of 5 distinct values of amplitude levels can make the simulation simpler. Therefore, isotropic hardening becomes the main objective function. This simulation's strain range is calculated using a uniaxial tensile test simulation result. Using the amplitude of the first, second, and third numerical loops, the isotropic hardening exponent is calibrated by graphical analysis. The stress amplitude does not fluctuate between specified levels but increases gradually and steadily within a defined range according to the load increase test (LIT). Thus, the segmentation of regions for distinct pairs is the sole vital point that has the greatest influence on the parameter estimation process. The curve fitting process delivers a satisfactory statistical and graphical result for AlSi10Mg and Ti-6Al-4V.

A more complex material characteristic can be achieved by more pairs of back stress parameters. Simulation efficiency, however, will suffer from overfitting. The bias in the simulated and calibrated parameters stems from the bias of the experimental data for both alloys. The calibration procedure for Ti-6Al-4V yields a considerably better outcome compared to that for AlSi10Mg. Simulation and experimental data are not similarly accurate in agreement in compression. The author recommends for future work to perform the calibration procedure on various strain rate levels or develop the applied cyclic plasticity model into a strain rate sensitive model. Segmentation and discretization of cyclic stress-strain response in the LIT were shown to be effective in producing load range-specific optimization of back stress pairs in AlSi10Mg and Ti-6Al-4V. One back stress cannot offer a near to the experiment data material calibration, but just a rough estimation of the material's cyclic characteristics. Having an over-defined material calibration is not more accurate and can lead to a nonefficient simulation. The goal of the mesh sensitivity research was to find the right number of elements to give outcomes equivalent to experimental values while reducing computation time. The roots of the notches at the defects are considered anyway to be locations of singularity. The values of stresses computed at these locations are just analyzed qualitatively. Adversely, the time it takes for computation grows exponentially when the mesh is overly refined. Isotropic hardening contributes to the development of plasticity until it reaches saturation. Cyclic hardening causes a decrease in overall strain. As a result, kinematic hardening plays a significant role in the development and build-up of creep plasticity beyond saturation till failure. The maximum total strain reaches a saturation level before continuing to increase until failure.

A representative volume element is suggested to capture such tiny microstructural deformation characteristics while still obtaining a viable converging FE model. A concept was presented for meso-microscale crack initiation time analysis, which makes use of crystallographic orientation data on various deposition planes to capture the anisotropy of additively manufactured materials. Unforced fatigue cracks that originate from slip bands on the maximum shear planes are likely to adopt a mode II shearing failure that will branch into a mode I tensile failure depending on homogeneities encountered along the crack's course. Fatemi-Socie damage parameter was considered as a material property that varies depending on the material and load situation. It can be connected to homogenized state variables resulting from the FE simulations. The ratio of stabilization of hysteresis loops was used to characterize the Fatemi-Socie damage parameter. The highest in-plane shear strain generated under simulated cyclic loading determines the fatigue lifetime prediction model based on Fatemi-Socie damage. The convergence of the employed plasticity model is more challenging due to small plastic strain increments. The damage parameter can then be expressed in terms of uniaxial loading utilizing the relationship models that have been established. In the HCF regime, the cyclic shear strain concentration factor inside the microstructure is greater than elastic loading. The critical plane corresponds to the minimum hydrostatic stress damage since the damage is mainly induced by the deviatoric load that is quantified by the T.F. Calculations relating to fatigue hot spots could then be used to determine the probability of fatigue fracture initiation in HCF and VHCF.

6.5   Bayesian Statistics

Fatigue is a random variable regardless of whether it is examined in the laboratory or on-site. The likelihood of a structure or component failing is a critical design parameter for structural analysis. Consideration of the stress range, stress level, and size effect is one of the most challenging and enticing problems that has not been completely resolved. Creating predictive models of fatigue behavior is always a great challenge in the technology of additive manufacturing due to the immense number of factors that can be extremely complex to physically incorporate into a deterministic modeling scheme. For a stress-controlled statistical fatigue lifetime model, a three-parameter Weibull distribution, such as in the case of fatigue strength in the S-N curves, was utilized, where the mean and variance of the probability density function were clearly defined. The stress-life model was derived based on the compatibility condition of life and load distributions. In a Woehler field, the compatibility of the cumulative density function (CDF) of the lifetime and the stress range must be ensured such that the minimization law of the weakest link principle is valid. The adopted concept comes in accordance with the idea that a fatigue limit for engineering metallic materials does not exist. There are several methods and procedures for the estimation of the Weibull and Gumbel parameters. For instance, the maximum likelihood estimation (MLE), which is widely used, was employed in this thesis. The weakness of the joint probability formulation is that it tends to maximize the likelihood, which results in an overshoot of the variance.

The shortcoming of the fatigue limit in the literature was redeemed in the presented procedure, where it was reported that the weakest link principle would be suitable only for a specific segment of the fatigue regimes, mostly the HCF. The joint probability function could be solved with a unique Jacobian matrix which is known as the Jacobian of the mapping, in case one fatigue-inducing factor has to be mapped on the fatigue lifetime. The formulation could be simplified if statistical independence can be established between the fatigue-related influences. Using Baye's rule, the expectation of certain marginal microstructural and porosity states in the continuous density functional case was established. The training data of numerical and experimental results formed the prior knowledge about fatigue and its related influences. Hence, the posterior distribution of fatigue-related influences could be calculated. A loss function measures the costs of making a statistical decision compared to the target fatigue lifetime. The defect-correlated assessment of fatigue strength was established using machine learning (ML) and Bayesian statistics algorithms. A contour plot of the related fatigue strength across the whole range of resulting defects is presented for a 50% probability of failure. The reduction in fatigue strength due to an increase in mean defect size is consistent across all studied fatigue decades based on the load density per unit volume. Therefore, the continuum damage mechanics (CDM) theory was efficiently integrated with ML models. The model was also used to integrate past information about microstructures and porosity, as well as uncertainties on fatigue life. Hence, mapping the fatigue strength of various applied energy densities was possible as a chart between energy density and load density per unit volume for the various fatigue regimes. The results show that failure can be identified before a technical fracture appears and that adjustments to the process can be made in time to avoid significant deterioration of properties. Overall, a novel technique has been developed that attempts to maintain a high degree of energy input while controlling cooling rates. Cumulative energy absorption increases as the power of energy function increases. In SLM, the higher energy density drives the Ti-6Al-4V powder bed to over-melt. It is planned in future studies following this thesis to include further effects such as residual stress mapping on fatigue strength.

6.6   Outlook

The Hall-Petch law predicts that AM produces fine-grained microstructures with exceptional strength. AM's complicated thermal cycle involves re-heating previously formed layers, resulting in an in situ heat treatment. This can lead to a breakdown of brittle phases into more ductile variants, alloying element partitioning, and grain development. High-performance materials such as titanium-aluminides, for example, are already being researched [407]. The comprehensive knowledge of the formation mechanisms for the observed microstructure, particle morphologies, and precipitates is ongoing. Heat treatments are being designed and associated static and dynamic mechanical characteristics evaluated. In-depth knowledge of lightweight alloys' processing principle might lead to the composition of additional AM alloys [408]. In future work, the effect of post-processing of AlSi10Mg alloys on the fatigue behavior should be examined using the second exposure treatment as well as the proposed machine learning algorithm. It is also important to investigate the relationship between oxide layer defects and fracture propagation, in situ, quasi-in situ, and ex situ experiments combining μ-CT and mechanical testing such as proposed in some literature [203].

Single-layer specimens with controlled lengths of scanning vectors should be manufactured to analyze the stability of the melt pool in relation to the vector length by observing the homogeneity of the specimen surface using in situ X-ray imaging as proposed to counter such potential risk of defect formation at scanning interfaces [192, 197]. A connection between vector length and melt pool stability should also be developed based on these first outcomes. Bulk specimens shall then be printed based on these findings, and the associated surface roughness must be evaluated using, e.g., laser confocal microscopy. It will be crucial to concentrate on the number of particles adhering to the surface which contributes to the final surface morphology. Analyzing the thermal conditions in the subsurface and surface region will reveal a link between temperature and the sintering or balling of particles to the surface. Understanding the movement of powder particles in the powder bed will be possible thanks to the use of a high-speed camera to monitor the operation. Therefore, the proposed future study will reveal the relationship between scanning vector length, processing parameters (melt pool stability and particle adhesion to the surface), and the resultant surface roughness. On the surface and subsurface, light microscopy, as well as μ-CT, can be used to study the structural morphology of the flaws and roughness caused by the suggested scanning techniques. To account for the differences in temperature conditions at various places on the specimen, measurements should be synchronized statistically at areas near the specimen surface, deposition platform, and free upper surfaces.

To induce certain microstructure and material properties, it is crucial to understand cooling conditions during solidification. Although processing settings can affect the cooling rate, other factors such as part shape and distance from the construction platform also have an impact. The goal of future research should be to figure out how total exposure time per layer affects thermal conditions, which leads to microstructure and residual stresses, which are influential for fatigue strength. Consequently, specimens from construction jobs with two to three levels of total exposure time per layer should be examined. Scanning electron microscopy and X-ray diffraction can show their microstructure characteristics and residual stresses, respectively. The change in thermal conditions can be tracked using a wide-field thermal camera and an on-axis photodiode. A connection between total exposure duration, cooling rate, and residual stresses should be developed based on the change in residual stresses and thermal data. As previously demonstrated, subsurface porosity forms at the intersection of contour and volume vectors [304]. The formation of subsurface pores is generally a factor in the production of keyholes leading to fatigue failure. The goal of future investigations is to learn more about how subsurface porosity forms. Also disclosed shall be the influence of laser power ramping volume vectors on flaw creation. It will be crucial to comprehend the impact of the lowest laser power as well as the power ramp's gradient. A thermal camera should be used to monitor the subsurface in detail; thus, fatigue lifetime distribution can be correlated not only with pore distribution as in this work but also with temperature condition and contour-volume vector arrangement using in-process measurement techniques. Scanning electron microscopy shall be used to study keyhole morphology at high magnification, as well as porosity below the 50 μm equivalent diameter revealed in this study, as well as energy-dispersive X-ray (EDX) measurements to see if non-homogeneity of composition is responsible for non-consistent viscosity, which causes the melt pool to collapse, according to the literature [191].

The understanding of the microstructure on this scale is enhanced by EBSD mapping of the melt pool boundaries, and in the center as well as melt track overlaps. The integration of scales enhances the understanding of the evolution of microstructure and subsequent properties, which enables the numerical analysis to build realistic and efficient representative models of the various scales under various loading conditions. Future microstructural investigations shall be carried out on pre- and post-mechanical testing specimens such that a comparative study between healthy and damaged microstructure is possible. The structured study will provide a comprehensive opportunity to investigate the response of SLM microstructures. Crack initiation and propagation mechanisms shall also be monitored for various scales of the structural domain. The crack formation mechanisms shall be analyzed to conclude the influence of graded structures on the formation of non-typical defects, which can be responsible for crack initiation. Using information obtained in the failure analysis, a scale integration is performed depending on crack propagation mechanisms, which are investigated during fractography investigations. The goal of this scale integration methodology is to obtain a better understanding of the transition from one fatigue cracking scale to the other on the fracture surfaces. The understanding makes structural health monitoring of components possible to calculate the remaining life based on damage tolerance mechanics and models presented in this thesis, as well as utilize the predictive power of the presented machine learning algorithm.

Numerous expansions are required in future numerical work to enhance and expand the conclusions of this thesis. The advancements include the introduction of the Ohno-Wang cyclic plasticity model, which decomposes the kinematic hardening rule, to better prediction of multiaxial ratcheting behavior via pore and notch defects with a multiaxial stress state. The second is an expansion of the Fatemi-Socie damage parameter that ensures convergence even over notch pathways by establishing a relationship between the stress intensity factor and the maximum in-plane shear strain. The third is a microstructure-sensitive model based on crystal slippage that enables modeling of cyclic deformation of multiple microstructures without having to redo the macroscale cyclic tests required to calibrate the macroscale model of the new microstructure. The primary objective is to determine the fatigue lifespan of additively constructed materials with arbitrary microstructures while taking damage processes into account throughout the short and long fracture propagation phases. Finally, this will be feasible without recalibration of the models when process parameters change due to the impossibility of repeating these strenuous studies for every potential microstructure through additive manufacturing. The Ohno-Wang model, in conjunction with the enhanced Fatemi-Socie damage parameter, enables a thorough calculation of fatigue lifespan that encompasses both phases. At the grain scale, the complete microstructure-sensitive model will depict and distinguish the first damage phase. The upshot of this example will be a robust model for predicting the fatigue lifespan of single alloying systems AlSi10Mg and Ti-6Al-4V that does not need recalibration.

Since the process chain in additive manufacturing technologies is challenging to control using physically deterministic techniques, a bigger emphasis should be made in the future on statistical and artificially intelligent methods. Therefore, a novel random regression called the hidden Markov tree (HMT) model [409], which is based on the least-squares support vector machine (LS-SVM) classification contourlet, will be utilized to classify structural data such as porosity and microstructural features resulting from microscopy and μ-CT. Following that, the Bayes shrink technique will be used to effectively filter all noisy contourlet HMT coefficients that resulted in the first classification steps. They are then fused together using the weighted average method to produce the estimates of biases and weights of the fatigue lifetime and fatigue loading values avoiding the shortcoming of normalizing a model on a fatigue endurance limit. The analysis will be conducted on various materials configurations contaminated with all sorts of random distributions of meso- and microstructural features, as well as numerical results that result from simulation of cyclic deformation and fracture mechanics. To functionalize overall fatigue characteristics, the optimal design aim for producing metals and alloys is to regulate microstructural features that contribute to fracture initiation and propagation resistance. It remains to be seen if current developments in computational materials science will result in the establishment of physically-based materials design principles based on artificial intelligence, which is capable of ultimately displacing century-old empirical methodologies.

References

1. Bremen, S., Meiners, W., Diatlov, A.: Selective laser melting: A manufacturing technology for the future? Laser Technik Journal. 9, 33–38 (2012). https://doi.org/10.1002/latj.201290018.
2. Gialanella, S., Malandruccolo, A.: Aerospace alloys. Springer International Publishing, Cham (2020). https://doi.org/10.1007/978-3-030-24440-8.
3. Cui, W.: A state-of-the-art review on fatigue life prediction methods for metal structures. Journal of Marine Science and Technology. 7, 43–56 (2002). https://doi.org/10.1007/s007730200012.
4. Johnson, N.S., Vulimiri, P.S., To, A.C., Zhang, X., Brice, C.A., Kappes, B.B., Stebner, A.P.: Invited review: Machine learning for materials developments in metals additive manufacturing. Additive Manufacturing. 36, 101641 (2020). https://doi.org/10.1016/j.addma.2020.101641.
5. Wang, C., Tan, X.P., Tor, S.B., Lim, C.S.: Machine learning in additive manufacturing: State-of-the-art and perspectives. Additive Manufacturing. 36, 101538 (2020). https://doi.org/10.1016/j.addma.2020.101538.
6. Lasi, H., Fettke, P., Kemper, H.-G., Feld, T., Hoffmann, M.: Industry 4.0. Business & Information Systems Engineering. 6, 239–242 (2014). https://doi.org/10.1007/s12599-014-0334-4.
7. Fayazfar, H., Salarian, M., Rogalsky, A., Sarker, D., Russo, P., Paserin, V., Toyserkani, E.: A critical review of powder-based additive manufacturing of ferrous alloys: Process parameters, microstructure and mechanical properties. Materials & Design. 144, 98–128 (2018). https://doi.org/10.1016/j.matdes.2018.02.018.
8. Singh, S., Ramakrishna, S., Singh, R.: Material issues in additive manufacturing: A review. Journal of Manufacturing Processes. 25, 185–200 (2017). https://doi.org/10.1016/j.jmapro.2016.11.006.
9. Zhang, D., Sun, S., Qiu, D., Gibson, M.A., Dargusch, M.S., Brandt, M., Qian, M., Easton, M.: Metal alloys for fusion-based additive manufacturing. Advanced Engineering Materials. 20, 1700952 (2018). https://doi.org/10.1002/adem.201700952.
10. Jiang, J., Xu, X., Stringer, J.: Support structures for additive manufacturing: A review. Journal of Manufacturing and Materials Processing. 2, 64 (2018). https://doi.org/10.3390/jmmp2040064.

© The Editor(s) (if applicable) and The Author(s), under exclusive license to Springer Fachmedien Wiesbaden GmbH, part of Springer Nature 2022
M. Mamduh Mustafa Awd, *Machine Learning Algorithm for Fatigue Fields in Additive Manufacturing*, Werkstofftechnische Berichte | Reports of Materials Science and Engineering, https:doi.org/10.1007/978-3-658-40237-2

219

11. Lewandowski, J.J., Seifi, M.: Metal additive manufacturing: A review of mechanical properties. Annual Review of Materials Research. 46, 151–186 (2016). https://doi.org/10.1146/annurev-matsci-070115-032024.
12. Everton, S.K., Hirsch, M., Stravroulakis, P., Leach, R.K., Clare, A.T.: Review of in situ process monitoring and in situ metrology for metal additive manufacturing. Materials & Design. 95, 431–445 (2016). https://doi.org/10.1016/j.matdes.2016.01.099.
13. Townsend, A., Senin, N., Blunt, L., Leach, R.K., Taylor, J.S.: Surface texture metrology for metal additive manufacturing: A review. Precision Engineering. 46, 34–47 (2016). https://doi.org/10.1016/j.precisioneng.2016.06.001.
14. Wong, K.V., Hernandez, A.: A review of additive manufacturing. ISRN Mechanical Engineering. 2012, 1–10 (2012). https://doi.org/10.5402/2012/208760.
15. Frazier, W.E.: Metal additive manufacturing: A review. Journal of Materials Engineering and Performance. 23, 1917–1928 (2014). https://doi.org/10.1007/s11665-014-0958-z.
16. Spierings, A.B., Levy, G., Wegener, K.: Designing material properties locally with additive manufacturing technology SLM. In: Solid Freeform Fabrication Symposium 2012, Austin, TX, USA, August 6–8, 2012. ETH Zurich. https://doi.org/10.3929/ethz-a-010335595.
17. Tapia, G., Elwany, A.: A review on process monitoring and control in metal-based additive manufacturing. Journal of Manufacturing Science and Engineering. 136, 060801 (2014). https://doi.org/10.1115/1.4028540.
18. Vaezi, M., Chianrabutra, S., Mellor, B., Yang, S.: Multiple material additive manufacturing—Part 1: A review. Virtual and Physical Prototyping. 8, 19–50 (2013). https://doi.org/10.1080/17452759.2013.778175.
19. Beretta, S., Romano, S.: A comparison of fatigue strength sensitivity to defects for materials manufactured by AM or traditional processes. International Journal of Fatigue. 94, 178–191 (2017). https://doi.org/10.1016/j.ijfatigue.2016.06.020.
20. Siddique, S., Imran, M., Walther, F.: Very high cycle fatigue and fatigue crack propagation behavior of selective laser melted AlSi12 alloy. International Journal of Fatigue. 94, 246–254 (2017). https://doi.org/10.1016/j.ijfatigue.2016.06.003.
21. Siddique, S., Awd, M., Tenkamp, J., Walther, F.: Development of a stochastic approach for fatigue life prediction of AlSi12 alloy processed by selective laser melting. Engineering Failure Analysis. 79, 34–50 (2017). https://doi.org/10.1016/j.engfailanal.2017.03.015.
22. Siddique, S., Imran, M., Wycisk, E., Emmelmann, C., Walther, F.: Fatigue assessment of laser additive manufactured AlSi12 eutectic alloy in the very high cycle fatigue (VHCF) range up to 1E9 cycles. Materials Today: Proceedings. 3, 2853–2860 (2016). https://doi.org/10.1016/j.matpr.2016.07.004.
23. Biffi, C.A., Fiocchi, J., Bassani, P., Paolino, D.S., Tridello, A., Chiandussi, G., Rossetto, M., Tuissi, A.: Microstructure and preliminary fatigue analysis on AlSi10Mg samples manufactured by SLM. Procedia Structural Integrity. 7, 50–57 (2017). https://doi.org/10.1016/j.prostr.2017.11.060.
24. Tang, M., Pistorius, P.C.: Fatigue life prediction for AlSi10Mg components produced by selective laser melting. International Journal of Fatigue. 125, 479–490 (2019). https://doi.org/10.1016/j.ijfatigue.2019.04.015.

25. Paolino, D.S., Tridello, A., Fiocchi, J., Biffi, C.A., Chiandussi, G., Rossetto, M., Tuissi, A.: VHCF response up to 1E9 cycles of SLM AlSi10Mg specimens built in a vertical direction. Applied Sciences. 9, 2954 (2019). https://doi.org/10.3390/app9152954.
26. Caivano, R., Tridello, A., Chiandussi, G., Qian, G., Paolino, D., Berto, F.: Very high cycle fatigue (VHCF) response of additively manufactured materials: A review. Fatigue & Fracture of Engineering Materials & Structures. 44, 2919–2943 (2021). https://doi.org/10.1111/ffe.13567.
27. Rao, J.H., Zhang, Y., Huang, A., Wu, X., Zhang, K.: Improving fatigue performances of selective laser melted Al-7Si-0.6Mg alloy via defects control. International Journal of Fatigue. 129, 105215 (2019). https://doi.org/10.1016/j.ijfatigue.2019.105215.
28. Tridello, A., Fiocchi, J., Biffi, C.A., Chiandussi, G., Rossetto, M., Tuissi, A., Paolino, D.S.: Influence of the annealing and defects on the VHCF behavior of an SLM AlSi10Mg alloy. Fatigue & Fracture of Engineering Materials & Structures. 42, 2794–2807 (2019). https://doi.org/10.1111/ffe.13123.
29. Bagherifard, S., Beretta, N., Monti, S., Riccio, M., Bandini, M., Guagliano, M.: On the fatigue strength enhancement of additive manufactured AlSi10Mg parts by mechanical and thermal post-processing. Materials & Design. 145, 28–41 (2018). https://doi.org/10.1016/j.matdes.2018.02.055.
30. Uzan, N.E., Ramati, S., Shneck, R., Frage, N., Yeheskel, O.: On the effect of shot-peening on fatigue resistance of AlSi10Mg specimens fabricated by additive manufacturing using selective laser melting (AM-SLM). Additive Manufacturing. 21, 458–464 (2018). https://doi.org/10.1016/j.addma.2018.03.030.
31. Tajiri, A., Nozaki, T., Uematsu, Y., Kakiuchi, T., Nakajima, M., Nakamura, Y., Tanaka, H.: Fatigue limit prediction of large scale cast aluminum alloy A356. Procedia Materials Science. 3, 924–929 (2014). https://doi.org/10.1016/j.mspro.2014.06.150.
32. Brandl, E., Heckenberger, U., Holzinger, V., Buchbinder, D.: Additive manufactured AlSi10Mg samples using selective laser melting (SLM): Microstructure, high cycle fatigue, and fracture behavior. Materials & Design. 34, 159–169 (2012). https://doi.org/10.1016/j.matdes.2011.07.067.
33. Di Giovanni, M.T., de Menezes, J.T.O., Bolelli, G., Cerri, E., Castrodeza, E.M.: Fatigue crack growth behavior of a selective laser melted AlSi10Mg. Engineering Fracture Mechanics. 217, 106564 (2019). https://doi.org/10.1016/j.engfracmech.2019.106564.
34. Romano, S., Brückner-Foit, A., Brandão, A., Gumpinger, J., Ghidini, T., Beretta, S.: Fatigue properties of AlSi10Mg obtained by additive manufacturing: Defect-based modelling and prediction of fatigue strength. Engineering Fracture Mechanics. 187, 165–189 (2018). https://doi.org/10.1016/j.engfracmech.2017.11.002.
35. Hu, Y.N., Wu, S.C., Wu, Z.K., Zhong, X.L., Ahmed, S., Karabal, S., Xiao, X.H., Zhang, H.O., Withers, P.J.: A new approach to correlate the defect population with the fatigue life of selective laser melted Ti-6Al-4V alloy. International Journal of Fatigue. 136, 105584 (2020). https://doi.org/10.1016/j.ijfatigue.2020.105584.
36. Zhao, J., Easton, M., Qian, M., Leary, M., Brandt, M.: Effect of building direction on porosity and fatigue life of selective laser melted AlSi12Mg alloy. Materials Science and Engineering: A. 729, 76–85 (2018). https://doi.org/10.1016/j.msea.2018.05.040.
37. Beevers, E., Brandão, A.D., Gumpinger, J., Gschweitl, M., Seyfert, C., Hofbauer, P., Rohr, T., Ghidini, T.: Fatigue properties and material characteristics of additively manufactured AlSi10Mg—Effect of the contour parameter on the microstructure, density, residual stress, roughness and mechanical properties. International Journal of Fatigue. 117, 148–162 (2018). https://doi.org/10.1016/j.ijfatigue.2018.08.023.

38. Le, V.-D., Saintier, N., Morel, F., Bellett, D., Osmond, P.: Investigation of the effect of porosity on the high cycle fatigue behaviour of cast Al-Si alloy by X-ray microtomography. International Journal of Fatigue. 106, 24–37 (2018). https://doi.org/10.1016/j.ijfatigue.2017.09.012.

39. Brandão, A.D., Gumpinger, J., Gschweitl, M., Seyfert, C., Hofbauer, P., Ghidini, T.: Fatigue properties of additively manufactured AlSi10Mg—Surface treatment effect. Procedia Structural Integrity. 7, 58–66 (2017). https://doi.org/10.1016/j.prostr.2017.11.061.

40. Zhang, C., Zhu, H., Liao, H., Cheng, Y., Hu, Z., Zeng, X.: Effect of heat treatments on fatigue property of selective laser melting AlSi10Mg. International Journal of Fatigue. 116, 513–522 (2018). https://doi.org/10.1016/j.ijfatigue.2018.07.016.

41. Kratochvilova, V., Vlasic, F., Mazal, P., Palousek, D.: Fatigue behaviour evaluation of additively and conventionally produced materials by acoustic emission method. Procedia Structural Integrity. 5, 393–400 (2017). https://doi.org/10.1016/j.prostr.2017.07.187.

42. Uzan, N.E., Shneck, R., Yeheskel, O., Frage, N.: Fatigue of AlSi10Mg specimens fabricated by additive manufacturing selective laser melting (AM-SLM). Materials Science and Engineering: A. 704, 229–237 (2017). https://doi.org/10.1016/j.msea.2017.08.027.

43. Aboulkhair, N.T., Maskery, I., Tuck, C., Ashcroft, I., Everitt, N.M.: Improving the fatigue behaviour of a selectively laser melted aluminium alloy: Influence of heat treatment and surface quality. Materials & Design. 104, 174–182 (2016). https://doi.org/10.1016/j.matdes.2016.05.041.

44. Greitemeier, D., Palm, F., Syassen, F., Melz, T.: Fatigue performance of additive manufactured TiAl6V4 using electron and laser beam melting. International Journal of Fatigue. 94, 211–217 (2017). https://doi.org/10.1016/j.ijfatigue.2016.05.001.

45. Gong, H., Nadimpalli, V.K., Rafi, K., Starr, T., Stucker, B.: Micro-CT evaluation of defects in Ti-6Al-4V parts fabricated by metal additive manufacturing. Technologies. 7, 44 (2019). https://doi.org/10.3390/technologies7020044.

46. Benedetti, M., Fontanari, V., Bandini, M., Zanini, F., Carmignato, S.: Low- and high-cycle fatigue resistance of Ti-6Al-4V ELI additively manufactured via selective laser melting: Mean stress and defect sensitivity. International Journal of Fatigue. 107, 96–109 (2018). https://doi.org/10.1016/j.ijfatigue.2017.10.021.

47. Biswal, R., Zhang, X., Shamir, M., Al Mamun, A., Awd, M., Walther, F., Khadar Syed, A.: Interrupted fatigue testing with periodic tomography to monitor porosity defects in wire+arc manufactured Ti-6Al-4V. Additive Manufacturing. 28, 517–527 (2019). https://doi.org/10.1016/j.addma.2019.04.026.

48. Walker, K.F., Liu, Q., Brandt, M.: Evaluation of fatigue crack propagation behaviour in Ti-6Al-4V manufactured by selective laser melting. International Journal of Fatigue. 104, 302–308 (2017). https://doi.org/10.1016/j.ijfatigue.2017.07.014.

49. Bagehorn, S., Wehr, J., Maier, H.J.: Application of mechanical surface finishing processes for roughness reduction and fatigue improvement of additively manufactured Ti-6Al-4V parts. International Journal of Fatigue. 102, 135–142 (2017). https://doi.org/10.1016/j.ijfatigue.2017.05.008.

50. Hrabe, N., Gnäupel-Herold, T., Quinn, T.: Fatigue properties of a titanium alloy (Ti–6Al–4V) fabricated via electron beam melting (EBM): Effects of internal defects and residual stress. International Journal of Fatigue. 94, 202–210 (2017). https://doi.org/10.1016/j.ijfatigue.2016.04.022.

51. Leuders, S., Meiners, S., Wu, L., Taube, A., Tröster, T., Niendorf, T.: Structural components manufactured by selective laser melting and investment casting—Impact of the process route on the damage mechanism under cyclic loading. Journal of Materials Processing Technology. 248, 130–142 (2017). https://doi.org/10.1016/j.jmatprotec.2017.04.026.

52. Benedetti, M., Torresani, E., Leoni, M., Fontanari, V., Bandini, M., Pederzolli, C., Potrich, C.: The effect of post-sintering treatments on the fatigue and biological behavior of Ti-6Al-4V ELI parts made by selective laser melting. Journal of the Mechanical Behavior of Biomedical Materials. 71, 295–306 (2017). https://doi.org/10.1016/j.jmbbm.2017.03.024.

53. Kasperovich, G., Hausmann, J.: Improvement of fatigue resistance and ductility of TiAl6V4 processed by selective laser melting. Journal of Materials Processing Technology. 220, 202–214 (2015). https://doi.org/10.1016/j.jmatprotec.2015.01.025.

54. Vayssette, B., Saintier, N., Brugger, C., Elmay, M., Pessard, E.: Surface roughness of Ti-6Al-4V parts obtained by SLM and EBM: Effect on the high cycle fatigue life. Procedia Engineering. 213, 89–97 (2018). https://doi.org/10.1016/j.proeng.2018.02.010.

55. Wycisk, E., Solbach, A., Siddique, S., Herzog, D., Walther, F., Emmelmann, C.: Effects of defects in laser additive manufactured Ti-6Al-4V on fatigue properties. Physics Procedia. 56, 371–378 (2014). https://doi.org/10.1016/j.phpro.2014.08.120.

56. Biswal, R., Syed, A.K., Zhang, X.: Assessment of the effect of isolated porosity defects on the fatigue performance of additive manufactured titanium alloy. Additive Manufacturing. 23, 433–442 (2018). https://doi.org/10.1016/j.addma.2018.08.024.

57. Sterling, A., Shamsaei, N., Torries, B., Thompson, S.M.: Fatigue behaviour of additively manufactured Ti-6Al-4V. Procedia Engineering. 133, 576–589 (2015). https://doi.org/10.1016/j.proeng.2015.12.632.

58. Leuders, S., Thöne, M., Riemer, A., Niendorf, T., Tröster, T., Richard, H.A., Maier, H.J.: On the mechanical behaviour of titanium alloy TiAl6V4 manufactured by selective laser melting: Fatigue resistance and crack growth performance. International Journal of Fatigue. 48, 300–307 (2013). https://doi.org/10.1016/j.ijfatigue.2012.11.011.

59. Scott-Emuakpor, O., Holycross, C., George, T., Knapp, K., Beck, J.: Fatigue and strength studies of Ti-6Al-4V fabricated by direct metal laser sintering. Journal of Engineering for Gas Turbines and Power. 138, 022101 (2016). https://doi.org/10.1115/1.4031271.

60. Wycisk, E., Siddique, S., Herzog, D., Walther, F., Emmelmann, C.: Fatigue performance of laser additive manufactured Ti-6Al-4V in very high cycle fatigue regime up to 1E9 cycles. Frontiers in Materials. 2, (2015). https://doi.org/10.3389/fmats.2015.00072.

61. Shui, X., Yamanaka, K., Mori, M., Nagata, Y., Kurita, K., Chiba, A.: Effects of postprocessing on cyclic fatigue response of a titanium alloy additively manufactured by electron beam melting. Materials Science and Engineering: A. 680, 239–248 (2017). https://doi.org/10.1016/j.msea.2016.10.059.

62. Saitova, L., Hoppel, H., Goken, M., Semenova, I., Valiev, R.: Cyclic deformation behavior and fatigue lives of ultrafine-grained Ti-6Al-4V ELI alloy for medical use. International Journal of Fatigue. 31, 322–331 (2009). https://doi.org/10.1016/j.ijfatigue.2008.08.007.

63. Zherebtsov, S., Salishchev, G., Galeyev, R., Maekawa, K.: Mechanical properties of Ti-6Al-4V titanium alloy with submicrocrystalline structure produced by severe plastic deformation. Materials Transactions. 46, 2020–2025 (2005). https://doi.org/10.2320/matertrans.46.2020.

64. Chan, K.S., Koike, M., Mason, R.L., Okabe, T.: Fatigue life of titanium alloys fabricated by additive layer manufacturing techniques for dental implants. Metallurgical and Materials Transactions A. 44, 1010–1022 (2013). https://doi.org/10.1007/s11661-012-1470-4.

65. Fatemi, A., Molaei, R., Sharifimehr, S., Phan, N., Shamsaei, N.: Multiaxial fatigue behavior of wrought and additive manufactured Ti-6Al-4V including surface finish effect. International Journal of Fatigue. 100, 347–366 (2017). https://doi.org/10.1016/j.ijfatigue.2017.03.044.

66. Wycisk, E., Emmelmann, C., Siddique, S., Walther, F.: High cycle fatigue (HCF) performance of Ti-6Al-4V alloy processed by selective laser melting. Advanced Materials Research. 816–817, 134–139 (2013). https://doi.org/10.4028/www.scientific.net/AMR.816-817.134.

67. Li, P., Warner, D.H., Fatemi, A., Phan, N.: Critical assessment of the fatigue performance of additively manufactured Ti–6Al–4V and perspective for future research. International Journal of Fatigue. 85, 130–143 (2016). https://doi.org/10.1016/j.ijfatigue.2015.12.003.

68. Günther, J., Krewerth, D., Lippmann, T., Leuders, S., Tröster, T., Weidner, A., Biermann, H., Niendorf, T.: Fatigue life of additively manufactured Ti–6Al–4V in the very high cycle fatigue regime. International Journal of Fatigue. 94, 236–245 (2017). https://doi.org/10.1016/j.ijfatigue.2016.05.018.

69. Zhai, Y., Galarraga, H., Lados, D.A.: Microstructure evolution, tensile properties, and fatigue damage mechanisms in Ti-6Al-4V alloys fabricated by two additive manufacturing techniques. Procedia Engineering. 114, 658–666 (2015). https://doi.org/10.1016/j.proeng.2015.08.007.

70. Sterling, A.J., Torries, B., Shamsaei, N., Thompson, S.M., Seely, D.W.: Fatigue behavior and failure mechanisms of direct laser deposited Ti–6Al–4V. Materials Science and Engineering: A. 655, 100–112 (2016). https://doi.org/10.1016/j.msea.2015.12.026.

71. Riemer, A., Richard, H.A.: Crack propagation in additive manufactured materials and structures. Procedia Structural Integrity. 2, 1229–1236 (2016). https://doi.org/10.1016/j.prostr.2016.06.157.

72. Liu, Y.J., Li, S.J., Wang, H.L., Hou, W.T., Hao, Y.L., Yang, R., Sercombe, T.B., Zhang, L.C.: Microstructure, defects and mechanical behavior of beta-type titanium porous structures manufactured by electron beam melting and selective laser melting. Acta Materialia. 113, 56–67 (2016). https://doi.org/10.1016/j.actamat.2016.04.029.

73. Chastand, V., Tezenas, A., Cadoret, Y., Quaegebeur, P., Maia, W., Charkaluk, E.: Fatigue characterization of titanium Ti-6Al-4V samples produced by additive manufacturing. Procedia Structural Integrity. 2, 3168–3176 (2016). https://doi.org/10.1016/j.prostr.2016.06.395.

74. Kahlin, M., Ansell, H., Moverare, J.J.: Fatigue behaviour of additive manufactured Ti6Al4V, with as-built surfaces, exposed to variable amplitude loading. International Journal of Fatigue. 103, 353–362 (2017). https://doi.org/10.1016/j.ijfatigue.2017.06.023.

75. Amin Yavari, S., Wauthle, R., van der Stok, J., Riemslag, A.C., Janssen, M., Mulier, M., Kruth, J.P., Schrooten, J., Weinans, H., Zadpoor, A.A.: Fatigue behavior of porous biomaterials manufactured using selective laser melting. Materials Science and Engineering: C. 33, 4849–4858 (2013). https://doi.org/10.1016/j.msec.2013.08.006.

76. Strantza, M., Van Hemelrijck, D., Guillaume, P., Aggelis, D.G.: Acoustic emission monitoring of crack propagation in additively manufactured and conventional titanium components. Mechanics Research Communications. 84, 8–13 (2017). https://doi.org/10.1016/j.mechrescom.2017.05.009.

77. Kahlin, M., Ansell, H., Moverare, J.J.: Fatigue behaviour of notched additive manufactured Ti6Al4V with as-built surfaces. International Journal of Fatigue. 101, 51–60 (2017). https://doi.org/10.1016/j.ijfatigue.2017.04.009.

78. Vrancken, B., Cain, V., Knutsen, R., Van Humbeeck, J.: Residual stress via the contour method in compact tension specimens produced via selective laser melting. Scripta Materialia. 87, 29–32 (2014). https://doi.org/10.1016/j.scriptamat.2014.05.016.

79. Rans, C., Michielssen, J., Walker, M., Wang, W., Hoen-Velterop, L. 't: Beyond the orthogonal: On the influence of build orientation on fatigue crack growth in SLM Ti-6Al-4V. International Journal of Fatigue. 116, 344–354 (2018). https://doi.org/10.1016/j.ijfatigue.2018.06.038.

80. Benedetti, M., Cazzolli, M., Fontanari, V., Leoni, M.: Fatigue limit of Ti6Al4V alloy produced by selective laser sintering. Procedia Structural Integrity. 2, 3158–3167 (2016). https://doi.org/10.1016/j.prostr.2016.06.394.

81. Seifi, M., Salem, A., Satko, D., Shaffer, J., Lewandowski, J.J.: Defect distribution and microstructure heterogeneity effects on fracture resistance and fatigue behavior of EBM Ti–6Al–4V. International Journal of Fatigue. 94, 263–287 (2017). https://doi.org/10.1016/j.ijfatigue.2016.06.001.

82. Zhai, Y., Galarraga, H., Lados, D.A.: Microstructure, static properties, and fatigue crack growth mechanisms in Ti-6Al-4V fabricated by additive manufacturing: LENS and EBM. Engineering Failure Analysis. 69, 3–14 (2016). https://doi.org/10.1016/j.engfailanal.2016.05.036.

83. Cain, V., Thijs, L., Van Humbeeck, J., Van Hooreweder, B., Knutsen, R.: Crack propagation and fracture toughness of Ti6Al4V alloy produced by selective laser melting. Additive Manufacturing. 5, 68–76 (2015). https://doi.org/10.1016/j.addma.2014.12.006.

84. Zhang, J., Zhang, X., Wang, X., Ding, J., Traoré, Y., Paddea, S., Williams, S.: Crack path selection at the interface of wrought and wire+arc additive manufactured Ti–6Al–4V. Materials & Design. 104, 365–375 (2016). https://doi.org/10.1016/j.matdes.2016.05.027.

85. Zhang, X., Syed, A.K., Biswal, R., Martina, F., Ding, J., Williams, S.: High cycle fatigue and fatigue crack growth rate in additive manufactured titanium alloys. In: Niepokolczycki, A. and Komorowski, J. (eds.) ICAF 2019—Structural Integrity in the Age of Additive Manufacturing. pp. 31–42. Springer International Publishing, Cham (2020). https://doi.org/10.1007/978-3-030-21503-3_3.

86. Tammas-Williams, S., Withers, P.J., Todd, I., Prangnell, P.B.: The influence of porosity on fatigue crack initiation in additively manufactured titanium components. Scientific Reports. 7, 7308 (2017). https://doi.org/10.1038/s41598-017-06504-5.

87. Jiao, Z.H., Xu, R.D., Yu, H.C., Wu, X.R.: Evaluation on tensile and fatigue crack growth performances of Ti6Al4V alloy produced by selective laser melting. Procedia Structural Integrity. 7, 124–132 (2017). https://doi.org/10.1016/j.prostr.2017.11.069.

88. Fatemi, A., Yang, L.: Cumulative fatigue damage and life prediction theories: A survey of the state of the art for homogeneous materials. International Journal of Fatigue. 20, 9–34 (1998). https://doi.org/10.1016/S0142-1123(97)00081-9.

89. Molent, L., Dixon, B.: Airframe metal fatigue revisited. International Journal of Fatigue. 131, 105323 (2020). https://doi.org/10.1016/j.ijfatigue.2019.105323.

90. Brusa, E., Sesana, R., Ossola, E.: Numerical modeling and testing of mechanical behavior of AM titanium alloy bracket for aerospace applications. Procedia Structural Integrity. 5, 753–760 (2017). https://doi.org/10.1016/j.prostr.2017.07.166.

91. Terriault, P., Brailovski, V.: Modeling and simulation of large, conformal, porosity-graded and lightweight lattice structures made by additive manufacturing. Finite Elements in Analysis and Design. 138, 1–11 (2018). https://doi.org/10.1016/j.finel.2017.09.005.

92. Gillner, K., Henrich, M., Münstermann, S.: Numerical study of inclusion parameters and their influence on fatigue lifetime. International Journal of Fatigue. 111, 70–80 (2018). https://doi.org/10.1016/j.ijfatigue.2018.01.036.

93. Yeratapally, S.R., Glavicic, M.G., Hardy, M., Sangid, M.D.: Microstructure based fatigue life prediction framework for polycrystalline nickel-base superalloys with emphasis on the role played by twin boundaries in crack initiation. Acta Materialia. 107, 152–167 (2016). https://doi.org/10.1016/j.actamat.2016.01.038.

94. Hedayati, R., Hosseini-Toudeshky, H., Sadighi, M., Mohammadi-Aghdam, M., Zadpoor, A.A.: Computational prediction of the fatigue behavior of additively manufactured porous metallic biomaterials. International Journal of Fatigue. 84, 67–79 (2016). https://doi.org/10.1016/j.ijfatigue.2015.11.017.

95. Zargarian, A., Esfahanian, M., Kadkhodapour, J., Ziaei-Rad, S.: Numerical simulation of the fatigue behavior of additive manufactured titanium porous lattice structures. Materials Science and Engineering: C. 60, 339–347 (2016). https://doi.org/10.1016/j.msec.2015.11.054.

96. Pellinghelli, D., Riboli, M., Spagnoli, A.: Full-model multiaxial fatigue life calculations with different criteria. Procedia Engineering. 213, 126–136 (2018). https://doi.org/10.1016/j.proeng.2018.02.014.

97. Meggiolaro, M.A., Castro, J.T.P. de, Wu, H.: Non-linear incremental fatigue damage calculation for multiaxial non-proportional histories. International Journal of Fatigue. 100, 502–511 (2017). https://doi.org/10.1016/j.ijfatigue.2016.12.008.

98. Tong, J., Zhao, L.G., Lin, B.: Ratchetting strain as a driving force for fatigue crack growth. International Journal of Fatigue. 46, 49–57 (2013). https://doi.org/10.1016/j.ijfatigue.2012.01.003.

99. Schnubel, D., Huber, N.: Retardation of fatigue crack growth in aircraft aluminium alloys via laser heating—Numerical prediction of fatigue crack growth. Computational Materials Science. 65, 461–469 (2012). https://doi.org/10.1016/j.commatsci.2012.07.047.

100. Bang, D.J., Ince, A., Noban, M.: Modeling approach for a unified crack growth model in short and long fatigue crack regimes. International Journal of Fatigue. 128, 105182 (2019). https://doi.org/10.1016/j.ijftigue.2019.06.042.

101. Kujawski, D.: A fatigue crack driving force parameter with load ratio effects. International Journal of Fatigue. 23, 239–246 (2001). https://doi.org/10.1016/S0142-112 3(01)00158-X.

102. El Haddad, M.H., Topper, T.H., Smith, K.N.: Prediction of non propagating cracks. Engineering Fracture Mechanics. 11, 573–584 (1979). https://doi.org/10.1016/0013-7944(79)90081-X.

103. Kujawski, D.: A new (ΔK+Kmax)0.5 driving force parameter for crack growth in aluminum alloys. International Journal of Fatigue. 23, 733–740 (2001). https://doi.org/10.1016/S0142-1123(01)00023-8.

104. Navarro, A., Rios, E.R.: A microstructurally-short fatigue crack growth equation. Fatigue & Fracture of Engineering Materials and Structures. 11, 383–396 (1988). https://doi.org/10.1111/j.1460-2695.1988.tb01391.x.

105. Donahue, R.J., Clark, H.M., Atanmo, P., Kumble, R., McEvily, A.J.: Crack opening displacement and the rate of fatigue crack growth. International Journal of Fracture Mechanics. 8, 209–219 (1972). https://doi.org/10.1007/BF00703882.

106. Wolf, E.: Fatigue crack closure under cyclic tension. Engineering Fracture Mechanics. 2, 37–45 (1970). https://doi.org/10.1016/0013-7944(70)90028-7.

107. Forman, R.G., Kearney, V.E., Engle, R.M.: Numerical analysis of crack propagation in cyclic-loaded structures. Journal of Basic Engineering. 89, 459–463 (1967). https://doi.org/10.1115/1.3609637.

108. Nasr Esfahani, S., Taheri Andani, M., Shayesteh Moghaddam, N., Mirzaeifar, R., Elahinia, M.: Independent tuning of stiffness and toughness of additively manufactured titanium-polymer composites: Simulation, fabrication, and experimental studies. Journal of Materials Processing Technology. 238, 22–29 (2016). https://doi.org/10.1016/j.jmatprotec.2016.06.035.

109. Carvalho, A.L.M., Martins, J.P., Voorlwad, H.J.C.: Fatigue damage accumulation in aluminum 7050-T7451 alloy subjected to block programs loading under step-down sequence. Procedia Engineering. 2, 2037–2043 (2010). https://doi.org/10.1016/j.proeng.2010.03.219.

110. Lu, Z., Xiang, Y., Liu, Y.: Crack growth-based fatigue-life prediction using an equivalent initial flaw model. Part II: Multiaxial loading. International Journal of Fatigue. 32, 376–381 (2010). https://doi.org/10.1016/j.ijfatigue.2009.07.013.

111. Glodež, S., Jezernik, N., Kramberger, J., Lassen, T.: Numerical modelling of fatigue crack initiation of martensitic steel. Advances in Engineering Software. 41, 823–829 (2010). https://doi.org/10.1016/j.advengsoft.2010.01.002.

112. Brucknerfoit, A., Huang, X.: Numerical simulation of micro-crack initiation of martensitic steel under fatigue loading. International Journal of Fatigue. 28, 963–971 (2006). https://doi.org/10.1016/j.ijfatigue.2005.08.011.

113. Jezernik, N., Kramberger, J., Lassen, T., Glodež, S.: Numerical modelling of fatigue crack initiation and growth of martensitic steels: Numerical simulation of micro-crack initiation. Fatigue & Fracture of Engineering Materials & Structures. no-no (2010). https://doi.org/10.1111/j.1460-2695.2010.01482.x.

114. Qian, J., Li, S.: Application of multiscale cohesive zone model to simulate fracture in polycrystalline solids. Journal of Engineering Materials and Technology. 133, 011010 (2011). https://doi.org/10.1115/1.4002647.

115. Santus, C., Taylor, D.: Physically short crack propagation in metals during high cycle fatigue. International Journal of Fatigue. 31, 1356–1365 (2009). https://doi.org/10.1016/j.ijfatigue.2009.03.002.
116. Fajdiga, G., Ren, Z., Kramar, J.: Comparison of virtual crack extension and strain energy density methods applied to contact surface crack growth. Engineering Fracture Mechanics. 74, 2721–2734 (2007). https://doi.org/10.1016/j.engfracmech.2007.01.016.
117. Huang, X., Brückner-Foit, A., Besel, M., Motoyashiki, Y.: Simplified three-dimensional model for fatigue crack initiation. Engineering Fracture Mechanics. 74, 2981–2991 (2007). https://doi.org/10.1016/j.engframech.2006.05.027.
118. Chapetti, M.: Fatigue propagation threshold of short cracks under constant amplitude loading. International Journal of Fatigue. 25, 1319–1326 (2003). https://doi.org/10.1016/S0142-1123(03)00065-3.
119. Jones, R., Molent, L., Walker, K.: Fatigue crack growth in a diverse range of materials. International Journal of Fatigue. 40, 43–50 (2012). https://doi.org/10.1016/j.ijfatigue.2012.01.004.
120. Kujawski, D.: Correlation of long- and physically short-cracks growth in aluminum alloys. Engineering Fracture Mechanics. 68, 1357–1369 (2001). https://doi.org/10.1016/S0013-7944(01)00029-7.
121. Xiulin, Z., Hirt, M.A.: Fatigue crack propagation in steels. Engineering Fracture Mechanics. 18, 965–973 (1983). https://doi.org/10.1016/0013-7944(83)90070-X.
122. McDowell, D.L., Dunne, F.P.E.: Microstructure-sensitive computational modeling of fatigue crack formation. International Journal of Fatigue. 32, 1521–1542 (2010). https://doi.org/10.1016/j.ijfatigue.2010.01.003.
123. R. G. Prasad, M., Biswas, A., Geenen, K., Amin, W., Gao, S., Lian, J., Röttger, A., Vajragupta, N., Hartmaier, A.: Influence of pore characteristics on anisotropic mechanical behavior of laser powder bed fusion–manufactured metal by micromechanical modeling. Advanced Engineering Materials. 22, 2000641 (2020). https://doi.org/10.1002/adem.202000641.
124. Bridier, F., McDowell, D.L., Villechaise, P., Mendez, J.: Crystal plasticity modeling of slip activity in Ti–6Al–4V under high cycle fatigue loading. International Journal of Plasticity. 25, 1066–1082 (2009). https://doi.org/10.1016/j.ijplas.2008.08.004.
125. Zhang, M., Zhang, J., McDowell, D.L.: Microstructure-based crystal plasticity modeling of cyclic deformation of Ti–6Al–4V. International Journal of Plasticity. 23, 1328–1348 (2007). https://doi.org/10.1016/j.ijplas.2006.11.009.
126. Mayeur, J.R., McDowell, D.L.: A three-dimensional crystal plasticity model for duplex Ti–6Al–4V. International Journal of Plasticity. 23, 1457–1485 (2007). https://doi.org/10.1016/j.ijplas.2006.11.006.
127. Sharaf, M., Kucharczyk, P., Vajragupta, N., Münstermann, S., Hartmaier, A., Bleck, W.: Modeling the microstructure influence on fatigue life variability in structural steels. Computational Materials Science. 94, 258–272 (2014). https://doi.org/10.1016/j.commatsci.2014.05.059.
128. Mlikota, M., Schmauder, S., Božić, Ž.: Calculation of the Wöhler (S-N) curve using a two-scale model. International Journal of Fatigue. 114, 289–297 (2018). https://doi.org/10.1016/j.ijfatigue.2018.03.018.

129. Tanaka, K., Mura, T.: A dislocation model for fatigue crack initiation. Journal of Applied Mechanics. 48, 97–103 (1981). https://doi.org/10.1115/1.3157599.
130. Baldissera, P., Delprete, C.: The formal analogy between Tanaka-Mura and Weibull models for high-cycle fatigue: Fatigue micro-mechanical models analogy. Fatigue & Fracture of Engineering Materials & Structures. 35, 114–121 (2012). https://doi.org/10.1111/j.1460-2695.2011.01598.x.
131. Mlikota, M., Schmauder, S., Božić, Ž., Hummel, M.: Modelling of overload effects on fatigue crack initiation in case of carbon steel: Modelling of overload effects on fatigue crack initiation. Fatigue & Fracture of Engineering Materials & Structures. 40, 1182–1190 (2017). https://doi.org/10.1111/ffe.12598.
132. Mlikota, M., Staib, S., Schmauder, S., Božić, Ž.: Numerical determination of Paris law constants for carbon steel using a two-scale model. Journal of Physics: Conference Series. 843, 012042 (2017). https://doi.org/10.1088/1742-6596/843/1/012042.
133. Božić, Ž., Schmauder, S., Mlikota, M., Hummel, M.: Multiscale fatigue crack growth modelling for welded stiffened panels: Multiscale fatigue crack growth modelling. Fatigue & Fracture of Engineering Materials & Structures. 37, 1043–1054 (2014). https://doi.org/10.1111/ffe.12189.
134. Gillner, K., Münstermann, S.: Numerically predicted high cycle fatigue properties through representative volume elements of the microstructure. International Journal of Fatigue. 105, 219–234 (2017). https://doi.org/10.1016/j.ijfatigue.2017.09.002.
135. Davis, B.R., Wawrzynek, P.A., Carter, B.J., Ingraffea, A.R.: 3D simulation of arbitrary crack growth using an energy-based formulation—Part II: Non-planar growth. Engineering Fracture Mechanics. 154, 111–127 (2016). https://doi.org/10.1016/j.engfracmech.2015.12.033.
136. Polák, J., Man, J.: Fatigue crack initiation—The role of point defects. International Journal of Fatigue. 65, 18–27 (2014). https://doi.org/10.1016/j.ijfatigue.2013.10.016.
137. Sistaninia, M., Niffenegger, M.: Prediction of damage-growth based fatigue life of polycrystalline materials using a microstructural modeling approach. International Journal of Fatigue. 66, 118–126 (2014). https://doi.org/10.1016/j.ijfatigue.2014.03.018.
138. Mikkola, E., Marquis, G., Solin, J.: Mesoscale modelling of crack nucleation from defects in steel. International Journal of Fatigue. 41, 64–71 (2012). https://doi.org/10.1016/j.ijfatigue.2011.12.022.
139. Vajragupta, N., Uthaisangsuk, V., Schmaling, B., Münstermann, S., Hartmaier, A., Bleck, W.: A micromechanical damage simulation of dual phase steels using XFEM. Computational Materials Science. 54, 271–279 (2012). https://doi.org/10.1016/j.com matsci.2011.10.035.
140. Lian, J., Yang, H., Vajragupta, N., Münstermann, S., Bleck, W.: A method to quantitatively upscale the damage initiation of dual-phase steels under various stress states from microscale to macroscale. Computational Materials Science. 94, 245–257 (2014). https://doi.org/10.1016/j.commatsci.2014.05.051.
141. Di Cicco, F., Fanelli, P., Vivio, F.: Fatigue reliability evaluation of riveted lap joints using a new rivet element and DFR. International Journal of Fatigue. 101, 430–438 (2017). https://doi.org/10.1016/j.ijfatigue.2017.02.006.

142. Wilson, D., Dunne, F.P.E.: A mechanistic modelling methodology for microstructure-sensitive fatigue crack growth. Journal of the Mechanics and Physics of Solids. 124, 827–848 (2019). https://doi.org/10.1016/j.jmps.2018.11.023.

143. Wilson, D., Wan, W., Dunne, F.P.E.: Microstructurally-sensitive fatigue crack growth in HCP, BCC and FCC polycrystals. Journal of the Mechanics and Physics of Solids. 126, 204–225 (2019). https://doi.org/10.1016/j.jmps.2019.02.012.

144. Barbosa, J.F., Correia, J.A., Júnior, R.F., Zhu, S.-P., Jesus, A.M.D.: Probabilistic S-N fields based on statistical distributions applied to metallic and composite materials: State of the art. Advances in Mechanical Engineering. 11, 1687814019870395 (2019). https://doi.org/10.1177/1687814019870395.

145. Correia, J., Apetre, N., Arcari, A., De Jesus, A., Muñiz-Calvente, M., Calçada, R., Berto, F., Fernández-Canteli, A.: Generalized probabilistic model allowing for various fatigue damage variables. International Journal of Fatigue. 100, 187–194 (2017). https://doi.org/10.1016/j.ijfatigue.2017.03.031.

146. Correia, J.A.F. de O., Pedrosa, B.A.S., Raposo, P.C., De Jesus, A.M.P., dos Santos Gervásio, H.M., Lesiuk, G.S., da Silva Rebelo, C.A., Calçada, R.A.B., da Silva, L.A.P.S.: Fatigue strength evaluation of resin-injected bolted connections using statistical analysis. Engineering. 3, 795–805 (2017). https://doi.org/10.1016/j.eng.2017.12.001.

147. Muñiz Calvente, M., Blasón, S., Fernández Canteli, A., de Jesús, A., Correia, J.: A probabilistic approach for multiaxial fatigue criteria. Frattura ed Integrità Strutturale. 11, 160–165 (2016). https://doi.org/10.3221/IGF-ESIS.39.16.

148. Liu, Y., Mahadevan, S.: Probabilistic fatigue life prediction using an equivalent initial flaw size distribution. International Journal of Fatigue. 31, 476–487 (2009). https://doi.org/10.1016/j.ijfatigue.2008.06.005.

149. Glodež, S., Šori, M., Kramberger, J.: A statistical evaluation of micro-crack initiation in thermally cut structural elements: A statistical evaluation of micro-crack initiation. Fatigue & Fracture of Engineering Materials & Structures. 36, 1298–1305 (2013). https://doi.org/10.1111/ffe.12068.

150. Hwang, C.G., Ingraffea, A.R.: Virtual crack extension method for calculating the second order derivatives of energy release rates for multiply cracked systems. Engineering Fracture Mechanics. 74, 1468–1487 (2007). https://doi.org/10.1016/j.engfracmech.2006.08.009.

151. Hoshide, T., Kusuura, K.: Life prediction by simulation of crack growth in notched components with different microstructures and under multiaxial fatigue. Fatigue & Fracture of Engineering Materials & Structures. 21, 201–213 (1998). https://doi.org/10.1046/j.1460-2695.1998.00492.x.

152. Glodež, S., Šori, M., Kramberger, J.: Prediction of micro-crack initiation in high strength steels using Weibull distribution. Engineering Fracture Mechanics. 108, 263–274 (2013). https://doi.org/10.1016/j.engfracmech.2013.02.015.

153. El Khoukhi, D., Morel, F., Saintier, N., Bellett, D., Osmond, P., Le, V.-D.: Probabilistic modeling of the size effect and scatter in high cycle fatigue using a Monte-Carlo approach: Role of the defect population in cast aluminum alloys. International Journal of Fatigue. 147, 106177 (2021). https://doi.org/10.1016/j.ijfatigue.2021.106177.

154. Zinovieva, O., Zinoviev, A., Ploshikhin, V.: Three-dimensional modeling of the microstructure evolution during metal additive manufacturing. Computational Materials Science. 141, 207–220 (2018). https://doi.org/10.1016/j.commatsci.2017.09.018.

155. Sahoo, S., Chou, K.: Phase-field simulation of microstructure evolution of Ti–6Al–4V in electron beam additive manufacturing process. Additive Manufacturing. 9, 14–24 (2016). https://doi.org/10.1016/j.addma.2015.12.005.

156. Yang, Q., Zhang, P., Cheng, L., Min, Z., Chyu, M., To, A.C.: Finite element modeling and validation of thermomechanical behavior of Ti-6Al-4V in directed energy deposition additive manufacturing. Additive Manufacturing. 12, 169–177 (2016). https://doi.org/10.1016/j.addma.2016.06.012.

157. Foteinopoulos, P., Papacharalampopoulos, A., Stavropoulos, P.: On thermal modeling of additive manufacturing processes. CIRP Journal of Manufacturing Science and Technology. 20, 66–83 (2018). https://doi.org/10.1016/j.cirpj.2017.09.007.

158. Conti, P., Cianetti, F., Pilerci, P.: Parametric finite elements model of SLM additive manufacturing process. Procedia Structural Integrity. 8, 410–421 (2018). https://doi.org/10.1016/j.prostr.2017.12.041.

159. Zhao, X., Iyer, A., Promoppatum, P., Yao, S.-C.: Numerical modeling of the thermal behavior and residual stress in the direct metal laser sintering process of titanium alloy products. Additive Manufacturing. 14, 126–136 (2017). https://doi.org/10.1016/j.addma.2016.10.005.

160. Denlinger, E.R., Michaleris, P.: Effect of stress relaxation on distortion in additive manufacturing process modeling. Additive Manufacturing. 12, 51–59 (2016). https://doi.org/10.1016/j.addma.2016.06.011.

161. Cao, J., Gharghouri, M.A., Nash, P.: Finite-element analysis and experimental validation of thermal residual stress and distortion in electron beam additive manufactured Ti-6Al-4V build plates. Journal of Materials Processing Technology. 237, 409–419 (2016). https://doi.org/10.1016/j.jmatprotec.2016.06.032.

162. Steuben, J.C., Iliopoulos, A.P., Michopoulos, J.G.: Discrete element modeling of particle-based additive manufacturing processes. Computer Methods in Applied Mechanics and Engineering. 305, 537–561 (2016). https://doi.org/10.1016/j.cma.2016.02.023.

163. Vastola, G., Zhang, G., Pei, Q.X., Zhang, Y.-W.: Modeling and control of remelting in high-energy beam additive manufacturing. Additive Manufacturing. 7, 57–63 (2015). https://doi.org/10.1016/j.addma.2014.12.004.

164. Li, Y., Gu, D.: Parametric analysis of thermal behavior during selective laser melting additive manufacturing of aluminum alloy powder. Materials & Design. 63, 856–867 (2014). https://doi.org/10.1016/j.matdes.2014.07.006.

165. Romano, J., Ladani, L., Sadowski, M.: Thermal modeling of laser based additive manufacturing processes within common materials. Procedia Manufacturing. 1, 238–250 (2015). https://doi.org/10.1016/j.promfg.2015.09.012.

166. Totten, G.E., MacKenzie, D.S. eds: Handbook of aluminum. M. Dekker, New York; Basel (2003).

167. Campbell, F.C. ed: Elements of metallurgy and engineering alloys. ASM International, Materials Park, Ohio (2008).

168. Davis, J.R., ASM International, ASM International eds: Properties and selection: Non-ferrous alloys and special-purpose materials. ASM International, Materials Park, Ohio (2000).

169. Lumley, R.N., TBX: Fundamentals of aluminium metallurgy. Elsevier Science (2011).

170. Polmear, I.J.: Light alloys: From traditional alloys to nanocrystals. Elsevier/Butterworth-Heinemann, Oxford; Burlington, MA (2006).

171. Kaufman, J.G.: Introduction to aluminum alloys and tempers. ASM International, Materials Park, OH (2000).

172. J. R. Davis & Associates, ASM International eds: Aluminum and aluminum alloys. ASM International, Materials Park, OH (1993).

173. Porter, D.A., Easterling, K.E., Sherif, M.Y.: Phase transformations in metals and alloys. CRC Press, Boca Raton, FL (2009).

174. Donachie, M.J., Donachie, S.J.: Superalloys: A technical guide. ASM International, Materials Park, OH (2002).

175. Leyens, C., Braun, R., Fröhlich, M., Hovsepian, P.Eh.: Recent progress in the coating protection of gamma titanium-aluminides. Journal of The Minerals, Metals & Materials Society (TMS). 58, 17–21 (2006). https://doi.org/10.1007/s11837-006-0062-4.

176. Neto, E.A. de S., Perić, D., Owen, D.R.J.: Computational methods for plasticity: Theory and applications. Wiley, Chichester, West Sussex, UK (2008).

177. Öchsner, A.: Continuum damage and fracture mechanics. Springer Singapore: Imprint: Springer, Singapore (2016). https://doi.org/10.1007/978-981-287-865-6.

178. Boyer, R.R.: Aerospace applications of beta titanium alloys. Journal of The Minerals, Metals & Materials Society (TMS). 46, 20–23 (1994). https://doi.org/10.1007/BF0322 0743.

179. Milewski, J.O.: Additive manufacturing of metals: From fundamental technology to rocket nozzles, medical implants, and custom jewelry. Springer International Publishing: Imprint: Springer, Cham (2017). https://doi.org/10.1007/978-3-319-58205-4.

180. Gibson, I., Rosen, D., Stucker, B.: Additive manufacturing technologies. Springer New York, New York, NY (2015). https://doi.org/10.1007/978-1-4939-2113-3.

181. Anandarajah, A.: Computational methods in elasticity and plasticity: Solids and porous media. Springer, New York (2010).

182. Mugwagwa, L., Dimitrov, D., Matope, S., Yadroitsev, I.: Influence of process parameters on residual stress related distortions in selective laser melting. Procedia Manufacturing. 21, 92–99 (2018). https://doi.org/10.1016/j.promfg.2018.02.099.

183. Awd, M., Johannsen, J., Chan, T., Merghany, M., Emmelmann, C., Walther, F.: Improvement of fatigue strength in lightweight selective laser melted alloys by in situ and ex situ composition and heat treatment. In: The Minerals, Metals & Materials Society (ed.) TMS 2020 149th Annual Meeting & Exhibition Supplemental Proceedings. pp. 115–126. Springer International Publishing, Cham (2020). https://doi.org/10.1007/978-3-030-36296-6_11.

184. Mall, S., Namjoshi, S.A., Porter, W.J.: Effects of microstructure on fretting fatigue crack initiation behavior of Ti-6Al-4V. Materials Science and Engineering: A. 383, 334–340 (2004). https://doi.org/10.1016/j.msea.2004.05.019.

185. DIN EN ISO 15708-3:2019–09, Zerstörungsfreie Prüfung_- Durchstrahlungsverfahren für Computertomographie_- Teil_3: Durchführung und Auswertung (ISO_15708-3:2017); Deutsche Fassung EN_ISO_15708-3:2019. Beuth Verlag GmbH. https://doi.org/10.31030/3054744.

186. Vogl, T.J., Clauß, W., Li, G.-Z., Yeon, K.M. eds: Computed tomography. Springer Berlin Heidelberg, Berlin, Heidelberg (1996). https://doi.org/10.1007/978-3-642-798 87-0.

187. De Chiffre, L., Carmignato, S., Kruth, J.-P., Schmitt, R., Weckenmann, A.: Industrial applications of computed tomography. CIRP Annals. 63, 655–677 (2014). https://doi. org/10.1016/j.cirp.2014.05.011.

188. Meyendorf, N.G.H., Nagy, P.B., Rokhlin, S.I. eds: Nondestructive materials characterization: With applications to aerospace materials. Springer Berlin Heidelberg, Berlin, Heidelberg (2004). https://doi.org/10.1007/978-3-662-08988-0.

189. Griffiths, P.R.: Beer's law. In: Chalmers, J.M. and Griffiths, P.R. (eds.) Handbook of vibrational spectroscopy. p. s4601. John Wiley & Sons, Ltd, Chichester, UK (2006). https://doi.org/10.1002/0470027320.s4601.

190. Kempen, K., Thijs, L., Van Humbeeck, J., Kruth, J.-P.: Processing AlSi10Mg by selective laser melting: Parameter optimisation and material characterisation. Materials Science and Technology. 31, 917–923 (2015). https://doi.org/10.1179/1743284714Y. 0000000702.

191. Awd, M., Siddique, S., Johannsen, J., Emmelmann, C., Walther, F.: Very high-cycle fatigue properties and microstructural damage mechanisms of selective laser melted AlSi10Mg alloy. International Journal of Fatigue. 124, 55–69 (2019). https://doi.org/ 10.1016/j.ijfatigue.2019.02.040.

192. Martin, A.A., Calta, N.P., Hammons, J.A., Khairallah, S.A., Nielsen, M.H., Shuttlesworth, R.M., Sinclair, N., Matthews, M.J., Jeffries, J.R., Willey, T.M., Lee, J.R.I.: Ultrafast dynamics of laser-metal interactions in additive manufacturing alloys captured by in situ X-ray imaging. Materials Today Advances. 1, 100002 (2019). https:// doi.org/10.1016/j.mtadv.2019.01.001.

193. Aboulkhair, N.T., Everitt, N.M., Ashcroft, I., Tuck, C.: Reducing porosity in AlSi10Mg parts processed by selective laser melting. Additive Manufacturing. 1–4, 77–86 (2014). https://doi.org/10.1016/j.addma.2014.08.001.

194. Read, N., Wang, W., Essa, K., Attallah, M.M.: Selective laser melting of AlSi10Mg alloy: Process optimisation and mechanical properties development. Materials & Design. 65, 417–424 (2015). https://doi.org/10.1016/j.matdes.2014.09.044.

195. Maskery, I., Aboulkhair, N.T., Corfield, M.R., Tuck, C., Clare, A.T., Leach, R.K., Wildman, R.D., Ashcroft, I.A., Hague, R.J.M.: Quantification and characterisation of porosity in selectively laser melted Al–Si10–Mg using X-ray computed tomography. Materials Characterization. 111, 193–204 (2016). https://doi.org/10.1016/j.mat char.2015.12.001.

196. Tammas-Williams, S., Withers, P.J., Todd, I., Prangnell, P.B.: Porosity regrowth during heat treatment of hot isostatically pressed additively manufactured titanium components. Scripta Materialia. 122, 72–76 (2016). https://doi.org/10.1016/j.scriptamat. 2016.05.002.

197. Martin, A.A., Calta, N.P., Khairallah, S.A., Wang, J., Depond, P.J., Fong, A.Y., Thampy, V., Guss, G.M., Kiss, A.M., Stone, K.H., Tassone, C.J., Nelson Weker, J., Toney, M.F., van Buuren, T., Matthews, M.J.: Dynamics of pore formation during laser powder bed fusion additive manufacturing. Nature Communications. 10, 1987 (2019). https://doi.org/10.1038/s41467-019-10009-2.

198. Yang, K.V., Rometsch, P., Jarvis, T., Rao, J., Cao, S., Davies, C., Wu, X.: Porosity formation mechanisms and fatigue response in Al-Si-Mg alloys made by selective laser melting. Materials Science and Engineering: A. 712, 166–174 (2018). https://doi.org/10.1016/j.msea.2017.11.078.

199. Kan, W.H., Nadot, Y., Foley, M., Ridosz, L., Proust, G., Cairney, J.M.: Factors that affect the properties of additively-manufactured AlSi10Mg: Porosity versus microstructure. Additive Manufacturing. 29, 100805 (2019). https://doi.org/10.1016/j.addma.2019.100805.

200. Martin, J.H., Yahata, B.D., Hundley, J.M., Mayer, J.A., Schaedler, T.A., Pollock, T.M.: 3D printing of high-strength aluminium alloys. Nature. 549, 365–369 (2017). https://doi.org/10.1038/nature23894.

201. Aboulkhair, N.T., Maskery, I., Tuck, C., Ashcroft, I., Everitt, N.M.: On the formation of AlSi10Mg single tracks and layers in selective laser melting: Microstructure and nano-mechanical properties. Journal of Materials Processing Technology. 230, 88–98 (2016). https://doi.org/10.1016/j.jmatprotec.2015.11.016.

202. Aboulkhair, N.T., Maskery, I., Tuck, C., Ashcroft, I., Everitt, N.M.: The microstructure and mechanical properties of selectively laser melted AlSi10Mg: The effect of a conventional T6-like heat treatment. Materials Science and Engineering: A. 667, 139–146 (2016). https://doi.org/10.1016/j.msea.2016.04.092.

203. Tradowsky, U., White, J., Ward, R.M., Read, N., Reimers, W., Attallah, M.M.: Selective laser melting of AlSi10Mg: Influence of post-processing on the microstructural and tensile properties development. Materials & Design. 105, 212–222 (2016). https://doi.org/10.1016/j.matdes.2016.05.066.

204. Weingarten, C., Buchbinder, D., Pirch, N., Meiners, W., Wissenbach, K., Poprawe, R.: Formation and reduction of hydrogen porosity during selective laser melting of AlSi10Mg. Journal of Materials Processing Technology. 221, 112–120 (2015). https://doi.org/10.1016/j.jmatprotec.2015.02.013.

205. Haubrich, J., Gussone, J., Barriobero-Vila, P., Kürnsteiner, P., Jägle, E.A., Raabe, D., Schell, N., Requena, G.: The role of lattice defects, element partitioning and intrinsic heat effects on the microstructure in selective laser melted Ti-6Al-4V. Acta Materialia. 167, 136–148 (2019). https://doi.org/10.1016/j.actamat.2019.01.039.

206. Ahn, Y.-K., Kim, H.-G., Park, H.-K., Kim, G.-H., Jung, K.-H., Lee, C.-W., Kim, W.-Y., Lim, S.-H., Lee, B.-S.: Mechanical and microstructural characteristics of commercial purity titanium implants fabricated by electron-beam additive manufacturing. Materials Letters. 187, 64–67 (2017). https://doi.org/10.1016/j.matlet.2016.10.064.

207. Zhang, X.-Y., Fang, G., Leeflang, S., Böttger, A.J., A. Zadpoor, A., Zhou, J.: Effect of subtransus heat treatment on the microstructure and mechanical properties of additively manufactured Ti-6Al-4V alloy. Journal of Alloys and Compounds. 735, 1562–1575 (2018). https://doi.org/10.1016/j.jallcom.2017.11.263.

208. Liu, Z., Lu, S.L., Tang, H.P., Qian, M., Zhan, L.: Characterization and decompositional crystallography of the massive phase grains in an additively-manufactured Ti-6Al-4V alloy. Materials Characterization. 127, 146–152 (2017). https://doi.org/10.1016/j.matchar.2017.01.012.

209. Agius, D., Kourousis, K.I., Wallbrink, C., Song, T.: Cyclic plasticity and microstructure of as-built SLM Ti-6Al-4V: The effect of build orientation. Materials Science and Engineering: A. 701, 85–100 (2017). https://doi.org/10.1016/j.msea.2017.06.069.

210. Hadadzadeh, A., Amirkhiz, B.S., Li, J., Odeshi, A., Mohammadi, M.: Deformation mechanism during dynamic loading of an additively manufactured AlSi10Mg_200C. Materials Science and Engineering: A. 722, 263–268 (2018). https://doi.org/10.1016/j.msea.2018.03.014.

211. Yang, K.V., Shi, Y., Palm, F., Wu, X., Rometsch, P.: Columnar to equiaxed transition in Al-Mg(-Sc)-Zr alloys produced by selective laser melting. Scripta Materialia. 145, 113–117 (2018). https://doi.org/10.1016/j.scriptamat.2017.10.021.

212. Skelton, R.P.: High temperature fatigue: Properties and prediction. Springer Netherlands, Dordrecht (1987).

213. Buhl, H.: Advanced aerospace materials. Springer, Berlin, Heidelberg (1992).

214. Tan, X., Kok, Y., Toh, W.Q., Tan, Y.J., Descoins, M., Mangelinck, D., Tor, S.B., Leong, K.F., Chua, C.K.: Revealing martensitic transformation and α/β interface evolution in electron beam melting three-dimensional-printed Ti-6Al-4V. Scientific Reports. 6, 26039 (2016). https://doi.org/10.1038/srep26039.

215. Krakhmalev, P., Fredriksson, G., Yadroitsava, I., Kazantseva, N., Plessis, A. du, Yadroitsev, I.: Deformation behavior and microstructure of Ti6Al4V manufactured by SLM. Physics Procedia. 83, 778–788 (2016). https://doi.org/10.1016/j.phpro.2016.08.080.

216. Xu, W., Brandt, M., Sun, S., Elambasseril, J., Liu, Q., Latham, K., Xia, K., Qian, M.: Additive manufacturing of strong and ductile Ti–6Al–4V by selective laser melting via in situ martensite decomposition. Acta Materialia. 85, 74–84 (2015). https://doi.org/10.1016/j.actamat.2014.11.028.

217. Vrancken, B., Thijs, L., Kruth, J.-P., Van Humbeeck, J.: Heat treatment of Ti-6Al-4V produced by selective laser melting: Microstructure and mechanical properties. Journal of Alloys and Compounds. 541, 177–185 (2012). https://doi.org/10.1016/j.jallcom.2012.07.022.

218. DIN EN ISO 6892-1:2020-06, Metallische Werkstoffe_- Zugversuch_- Teil_1: Prüfverfahren bei Raumtemperatur (ISO_6892-1:2019); Deutsche Fassung EN_ISO_6892-1:2019. Beuth Verlag GmbH. https://doi.org/10.31030/3132591.

219. DIN EN ISO 9513:2013-05, Metallische Werkstoffe_- Kalibrierung von Längenänderungs-Messeinrichtungen für die Prüfung mit einachsiger Beanspruchung (ISO_9513:2012_+_Cor._1:2013); Deutsche Fassung EN_ISO_9513:2012. Beuth Verlag GmbH. https://doi.org/10.31030/1912742.

220. Mohamed, A.M.A., Samuel, A.M., Samuel, F.H., Doty, H.W.: Influence of additives on the microstructure and tensile properties of near-eutectic Al–10.8%Si cast alloy. Materials & Design. 30, 3943–3957 (2009). https://doi.org/10.1016/j.matdes.2009.05.042.

221. Kempen, K., Thijs, L., Van Humbeeck, J., Kruth, J.-P.: Mechanical properties of AlSi10Mg produced by selective laser melting. Physics Procedia. 39, 439–446 (2012). https://doi.org/10.1016/j.phpro.2012.10.059.

222. Tang, M., Pistorius, P.C.: Oxides, porosity and fatigue performance of AlSi10Mg parts produced by selective laser melting. International Journal of Fatigue. 94, 192–201 (2017). https://doi.org/10.1016/j.ijfatigue.2016.06.002.

223. Casati, R., Hamidi Nasab, M., Coduri, M., Tirelli, V., Vedani, M.: Effects of platform pre-heating and thermal-treatment strategies on properties of AlSi10Mg alloy processed by selective laser melting. Metals. 8, 954 (2018). https://doi.org/10.3390/met8110954.

224. Sridharan, N., Gussev, M., Seibert, R., Parish, C., Norfolk, M., Terrani, K., Babu, S.S.: Rationalization of anisotropic mechanical properties of Al-6061 fabricated using ultrasonic additive manufacturing. Acta Materialia. 117, 228–237 (2016). https://doi.org/10.1016/j.actamat.2016.06.048.

225. Murr, L.E., Esquivel, E.V., Quinones, S.A., Gaytan, S.M., Lopez, M.I., Martinez, E.Y., Medina, F., Hernandez, D.H., Martinez, E., Martinez, J.L., Stafford, S.W., Brown, D.K., Hoppe, T., Meyers, W., Lindhe, U., Wicker, R.B.: Microstructures and mechanical properties of electron beam-rapid manufactured Ti–6Al–4V biomedical prototypes compared to wrought Ti–6Al–4V. Materials Characterization. 60, 96–105 (2009). https://doi.org/10.1016/j.matchar.2008.07.006.

226. Qiu, C., Adkins, N.J.E., Attallah, M.M.: Microstructure and tensile properties of selectively laser-melted and of HIPed laser-melted Ti–6Al–4V. Materials Science and Engineering: A. 578, 230–239 (2013). https://doi.org/10.1016/j.msea.2013.04.099.

227. Simonelli, M., Tse, Y.Y., Tuck, C.: Effect of the build orientation on the mechanical properties and fracture modes of SLM Ti–6Al–4V. Materials Science and Engineering: A. 616, 1–11 (2014). https://doi.org/10.1016/j.msea.2014.07.086.

228. Kang, N., Coddet, P., Chen, C., Wang, Y., Liao, H., Coddet, C.: Microstructure and wear behavior of in-situ hypereutectic Al–high Si alloys produced by selective laser melting. Materials & Design. 99, 120–126 (2016). https://doi.org/10.1016/j.matdes.2016.03.053.

229. Aversa, A., Lorusso, M., Cattano, G., Manfredi, D., Calignano, F., Ambrosio, E.P., Biamino, S., Fino, P., Lombardi, M., Pavese, M.: A study of the microstructure and the mechanical properties of an Al Si Ni alloy produced via selective laser melting. Journal of Alloys and Compounds. 695, 1470–1478 (2017). https://doi.org/10.1016/j.jallcom.2016.10.285.

230. Montero-Sistiaga, M.L., Mertens, R., Vrancken, B., Wang, X., Van Hooreweder, B., Kruth, J.-P., Van Humbeeck, J.: Changing the alloy composition of Al7075 for better processability by selective laser melting. Journal of Materials Processing Technology. 238, 437–445 (2016). https://doi.org/10.1016/j.jmatprotec.2016.08.003.

231. Song, B., Dong, S., Zhang, B., Liao, H., Coddet, C.: Effects of processing parameters on microstructure and mechanical property of selective laser melted Ti-6Al-4V. Materials & Design. 35, 120–125 (2012). https://doi.org/10.1016/j.matdes.2011.09.051.

232. Brandl, E., Schoberth, A., Leyens, C.: Morphology, microstructure, and hardness of titanium (Ti-6Al-4V) blocks deposited by wire-feed additive layer manufacturing (ALM). Materials Science and Engineering: A. 532, 295–307 (2012). https://doi.org/10.1016/j.msea.2011.10.095.

233. Gong, H., Rafi, K., Gu, H., Janaki Ram, G.D., Starr, T., Stucker, B.: Influence of defects on mechanical properties of Ti–6Al–4V components produced by selective laser melting and electron beam melting. Materials & Design. 86, 545–554 (2015). https://doi.org/10.1016/j.matdes.2015.07.147.

234. Shunmugavel, M., Polishetty, A., Littlefair, G.: Microstructure and mechanical properties of wrought and additive manufactured Ti-6Al-4V cylindrical bars. Procedia Technology. 20, 231–236 (2015). https://doi.org/10.1016/j.protcy.2015.07.037.

235. Mahamood, R.M., Akinlabi, E.T.: Scanning speed influence on the microstructure and micro hardness properties of titanium alloy produced by laser metal deposition process. Materials Today: Proceedings. 4, 5206–5214 (2017). https://doi.org/10.1016/j.matpr.2017.05.028.

236. Zhang, F., Yang, M., Clare, A.T., Lin, X., Tan, H., Chen, Y.: Microstructure and mechanical properties of Ti-2Al alloyed with Mo formed in laser additive manufacture. Journal of Alloys and Compounds. 727, 821–831 (2017). https://doi.org/10.1016/j.jallcom.2017.07.324.

237. Cai, C., Radoslaw, C., Zhang, J., Yan, Q., Wen, S., Song, B., Shi, Y.: In-situ preparation and formation of TiB/Ti-6Al-4V nanocomposite via laser additive manufacturing: Microstructure evolution and tribological behavior. Powder Technology. 342, 73–84 (2019). https://doi.org/10.1016/j.powtec.2018.09.088.

238. Doerner, M.F., Gardner, D.S., Nix, W.D.: Plastic properties of thin films on substrates as measured by submicron indentation hardness and substrate curvature techniques. Journal of Materials Research. 1, 845–851 (1986). https://doi.org/10.1557/JMR.1986.0845.

239. Pharr, G.M., Oliver, W.C.: Measurement of thin film mechanical properties using nanoindentation. MRS Bulletin. 17, 28–33 (1992). https://doi.org/10.1557/S0883769400041634.

240. Burnett, P.J., Rickerby, D.S.: The mechanical properties of wear-resistant coatings. Thin Solid Films. 148, 51–65 (1987). https://doi.org/10.1016/0040-6090(87)90120-9.

241. Sneddon, I.N.: The relation between load and penetration in the axisymmetric boussinesq problem for a punch of arbitrary profile. International Journal of Engineering Science. 3, 47–57 (1965). https://doi.org/10.1016/0020-7225(65)90019-4.

242. Tabor, D.: The hardness of solids. Review of Physics in Technology. 1, 145–179 (1970). https://doi.org/10.1088/0034-6683/1/3/i01.

243. Kushch, V.I., Dub, S.N., Litvin, P.M.: Determination of the young modulus from elastic section of the Berkovich indenter loading curve. Journal of Superhard Materials. 29, 228–234 (2007). https://doi.org/10.3103/S1063457607040065.

244. Oliver, W.C., Pharr, G.M.: An improved technique for determining hardness and elastic modulus using load and displacement sensing indentation experiments. Journal of Materials Research. 7, 1564–1583 (1992). https://doi.org/10.1557/JMR.1992.1564.

245. Oliver, W.C., Pharr, G.M.: Measurement of hardness and elastic modulus by instrumented indentation: Advances in understanding and refinements to methodology. Journal of Materials Research. 19, 3–20 (2004). https://doi.org/10.1557/jmr.2004.19.1.3.

246. Lucas, B.N., Oliver, W.C., Pharr, G.M., Loubet, J.-L.: Time dependent deformation during indentation testing. MRS Proceedings. 436, 233 (1996). https://doi.org/10.1557/PROC-436-233.

247. Thurn, J., Cook, R.F.: Simplified area function for sharp indenter tips in depth-sensing indentation. Journal of Materials Research. 17, 1143–1146 (2002). https://doi.org/10.1557/JMR.2002.0169.

248. Troyon, M., Huang, L.: Comparison of different analysis methods in nanoindentation and influence on the correction factor for contact area. Surface and Coatings Technology. 201, 1613–1619 (2006). https://doi.org/10.1016/j.surfcoat.2006.02.033.

249. Bei, H., George, E.P., Hay, J.L., Pharr, G.M.: Influence of indenter tip geometry on elastic deformation during nanoindentation. Physical Review Letters. 95, 045501 (2005). https://doi.org/10.1103/PhysRevLett.95.045501.

250. Antunes, J.M., Cavaleiro, A., Menezes, L.F., Simões, M.I., Fernandes, J.V.: Ultramicrohardness testing procedure with Vickers indenter. Surface and Coatings Technology. 149, 27–35 (2002). https://doi.org/10.1016/S0257-8972(01)01413-X.

251. Chicot, D., Yetna N'Jock, M., Puchi-Cabrera, E.S., Iost, A., Staia, M.H., Louis, G., Bouscarrat, G., Aumaitre, R.: A contact area function for Berkovich nanoindentation: Application to hardness determination of a TiHfCN thin film. Thin Solid Films. 558, 259–266 (2014). https://doi.org/10.1016/j.tsf.2014.02.044.

252. Schneider-Maunoury, C., Albayda, A., Bartier, O., Weiss, L., Mauvoisin, G., Hernot, X., Laheurte, P.: On the use of instrumented indentation to characterize the mechanical properties of functionally graded binary alloys manufactured by additive manufacturing. Materials Today Communications. 25, 101451 (2020). https://doi.org/10.1016/j.mtcomm.2020.101451.

253. Lu, L., Dao, M., Kumar, P., Ramamurty, U., Karniadakis, G.E., Suresh, S.: Extraction of mechanical properties of materials through deep learning from instrumented indentation. Proceedings of the National Academy of Sciences. 117, 7052–7062 (2020). https://doi.org/10.1073/pnas.1922210117.

254. Kese, K.O., Alvarez, A.-M., Karlsson, J.K.-H., Nilsson, K.H.: Experimental evaluation of nanoindentation as a technique for measuring the hardness and Young's modulus of Zircaloy-2 sheet material. Journal of Nuclear Materials. 507, 267–275 (2018). https://doi.org/10.1016/j.jnucmat.2018.04.029.

255. Siddique, S.: Reliability of selective laser melted AlSi12 alloy for quasistatic and fatigue applications. Springer Fachmedien Wiesbaden, Wiesbaden (2019). https://doi.org/10.1007/978-3-658-23425-6.

256. Ellyin, F.: Fatigue damage, crack growth and life prediction. Springer, Place of publication not identified (2012).

257. Wang, Q.G., Apelian, D., Lados, D.A.: Fatigue behavior of A356-T6 aluminum cast alloys. Part I: Effect of casting defects. Journal of Light Metals. 1, 73–84 (2001). https://doi.org/10.1016/S1471-5317(00)00008-0.

258. Tridello, A., Biffi, C.A., Fiocchi, J., Bassani, P., Chiandussi, G., Rossetto, M., Tuissi, A., Paolino, D.S.: VHCF response of as-built SLM AlSi10Mg specimens with large loaded volume. Fatigue & Fracture of Engineering Materials & Structures. 41, 1918–1928 (2018). https://doi.org/10.1111/ffe.12830.

259. Domfang Ngnekou, J.N., Nadot, Y., Henaff, G., Nicolai, J., Ridosz, L.: Influence of defect size on the fatigue resistance of AlSi10Mg alloy elaborated by selective laser melting (SLM). Procedia Structural Integrity. 7, 75–83 (2017). https://doi.org/10.1016/j.prostr.2017.11.063.

260. Domfang Ngnekou, J.N., Nadot, Y., Henaff, G., Nicolai, J., Kan, W.H., Cairney, J.M., Ridosz, L.: Fatigue properties of AlSi10Mg produced by additive layer manufacturing. International Journal of Fatigue. 119, 160–172 (2019). https://doi.org/10.1016/j.ijfatigue.2018.09.029.

261. Ngnekou, J.N.D., Henaff, G., Nadot, Y., Nicolai, J., Ridosz, L.: Fatigue resistance of selectively laser melted aluminum alloy under T6 heat treatment. Procedia Engineering. 213, 79–88 (2018). https://doi.org/10.1016/j.proeng.2018.02.009.

262. Xie, Y., Gao, M., Wang, F., Zhang, C., Hao, K., Wang, H., Zeng, X.: Anisotropy of fatigue crack growth in wire arc additive manufactured Ti-6Al-4V. Materials Science and Engineering: A. 709, 265–269 (2018). https://doi.org/10.1016/j.msea.2017.10.064.

263. Benenson, W. ed: Handbook of physics. Springer, New York (2001).

264. Schijve, J.: Fatigue of structures and materials. Springer, Dordrecht (2009).

265. Ebara, R.: The present situation and future problems in ultrasonic fatigue testing—Mainly reviewed on environmental effects and materials' screening. International Journal of Fatigue. 28, 1465–1470 (2006). https://doi.org/10.1016/j.ijfatigue.2005.04.019.

266. Morrissey, R., McDowell, D.L., Nicholas, T.: Frequency and stress ratio effects in high cycle fatigue of Ti-6Al-4V. International Journal of Fatigue. 21, 679–685 (1999). https://doi.org/10.1016/S0142-1123(99)00030-4.

267. Stanzl-Tschegg, S.E., Mayer, H.: Fatigue and fatigue crack growth of aluminium alloys at very high numbers of cycles. International Journal of Fatigue. 23, 231–237 (2001). https://doi.org/10.1016/S0142-1123(01)00167-0.

268. Wu, R., Zhang, D., Yu, Q., Jiang, Y., Arola, D.: Health monitoring of wind turbine blades in operation using three-dimensional digital image correlation. Mechanical Systems and Signal Processing. 130, 470–483 (2019). https://doi.org/10.1016/j.ymssp.2019.05.031.

269. Janssen, M., Zuidema, J., Wanhill, R.J.H.: Fracture mechanics. Spon Press, London; New York (2004).

270. Gdoutos, E.E.: Linear elastic stress field in cracked bodies. In: Fracture Mechanics Criteria and Applications. pp. 15–75. Springer Netherlands, Dordrecht (1990). https://doi.org/10.1007/978-94-009-1956-3_2.

271. Anderson, T.L.: Fracture mechanics: Fundamentals and applications. CRC Press/Taylor & Francis, Boca Raton (2017).

272. Hellen, T.: How to undertake fracture mechanics analysis with finite elements. NAFEMS, the International Association for the Engineering Analysis Community, Glasgow (2001).

273. Rice, J.R.: A path independent integral and the approximate analysis of strain concentration by notches and cracks. Journal of Applied Mechanics. 35, 379–386 (1968). https://doi.org/10.1115/1.3601206.

274. Angela, D.D., Ercolino, M.: Finite element analysis of fatigue response of nickel steel compact tension samples using Abaqus. Procedia Structural Integrity. 13, 939–946 (2018). https://doi.org/10.1016/j.prostr.2018.12.176.

275. Singh, I.V., Mishra, B.K., Bhattacharya, S., Patil, R.U.: The numerical simulation of fatigue crack growth using extended finite element method. International Journal of Fatigue. 36, 109–119 (2012). https://doi.org/10.1016/j.ijfatigue.2011.08.010.

276. Miranda, A.C.O., Meggiolaro, M.A., Castro, J.T.P., Martha, L.F., Bittencourt, T.N.: Fatigue life and crack path predictions in generic 2D structural components. Engineering Fracture Mechanics. 70, 1259–1279 (2003). https://doi.org/10.1016/S0013-7944(02)00099-1.

277. Erdogan, F., Sih, G.C.: Discussion on the crack extension in plates under plane loading and transverse shear. Journal of Basic Engineering. 85, 527–527 (1963). https://doi.org/10.1115/1.3656899.

278. Cotterell, B., Rice, J.R.: Slightly curved or kinked cracks. International Journal of Fracture. 16, 155–169 (1980). https://doi.org/10.1007/BF00012619.

279. Sghayer, A., Sedmak, A., Dinulović, M., Grozdanović, I.: Experimental and numerical analysis of fatigue crack growth in integral skin-stringer panels. Tehnicki vjesnik - Technical Gazette. 25, (2018). https://doi.org/10.17559/TV-20170308110329.

280. Mabson, G., Deobald, L., Dopker, B., Hoyt, D., Baylor, J., Graesser, D.: Fracture inter-
     face elements for static and fatigue analysis. In: 16th International Conference On
     Composite Materials (ICCM 16) (2007).
281. E08 Committee: Test method for measurement of fatigue crack growth rates. ASTM
     International. https://doi.org/10.1520/E0647-15E01.
282. Carpinteri, A., Paggi, M.: Are the Paris' law parameters dependent on each other?
     Frattura ed Integrità Strutturale. 1, 10–16 (2008). https://doi.org/10.3221/IGF-ESIS.
     02.02.
283. Melson, J.: Fatigue crack growth analysis with finite element methods and a Monte
     Carlo simulation, https://vtechworks.lib.vt.edu/handle/10919/48432, (2014).
284. Zamani, A., Eslami, M.R.: Implementation of the extended finite element method for
     dynamic thermoelastic fracture initiation. International Journal of Solids and Struc-
     tures. 47, 1392–1404 (2010). https://doi.org/10.1016/j.ijsolstr.2010.01.024.
285. Belytschko, T., Black, T.: Elastic crack growth in finite elements with minimal
     remeshing. International Journal for Numerical Methods in Engineering. 45, 601–620
     (1999). https://doi.org/10.1002/(SICI)1097-0207(19990620)45:5%3C601::AID-NME
     598%3E3.0.CO;2-S.
286. Xin, H., Correia, J.A.F.O., Veljkovic, M.: Three-dimensional fatigue crack propagation
     simulation using extended finite element methods for steel grades S355 and S690 con-
     sidering mean stress effects. Engineering Structures. 227, 111414 (2021). https://doi.
     org/10.1016/j.engstruct.2020.111414.
287. He, H., Liu, H., Zhu, C., Mura, A.: Numerical study on fatigue crack prop-
     agation behaviors in lubricated rolling contact. Chinese Journal of Aeronautics.
     S1000936121001035 (2021). https://doi.org/10.1016/j.cja.2021.03.012.
288. Hedayati, E., Vahedi, M.: Using extended finite element method for computation of the
     stress intensity factor, crack growth simulation and predicting fatigue crack growth in
     a slant-cracked plate of 6061-T651 aluminum. World Journal of Mechanics. 04, 24–30
     (2014). https://doi.org/10.4236/wjm.2014.41003.
289. Lee, S.H., Jeon, I.: 3D analysis of crack behavior using XFEM. Applied Mechanics
     and Materials. 789–790, 278–281 (2015). https://doi.org/10.4028/www.scientific.net/
     AMM.789-790.278.
290. Shi, J., Chopp, D., Lua, J., Sukumar, N., Belytschko, T.: Abaqus implementation of
     extended finite element method using a level set representation for three-dimensional
     fatigue crack growth and life predictions. Engineering Fracture Mechanics. 77, 2840–
     2863 (2010). https://doi.org/10.1016/j.engfracmech.2010.06.009.
291. Nittur, P.G., Karlsson, A.M., Carlsson, L.A.: Numerical evaluation of Paris-regime
     crack growth rate based on plastically dissipated energy. Engineering Fracture Mechan-
     ics. 124–125, 155–166 (2014). https://doi.org/10.1016/j.engfracmech.2014.04.013.
292. Goqo, S.: Computational study of compact tension and double torsion test geometries,
     http://hdl.handle.net/11427/9108, (2014).
293. Butnariu, D., Censor, Y., Reich, S. eds: Inherently parallel algorithms in feasibility and
     optimization and their applications. Elsevier, Amsterdam (2001).
294. Cherepanov, G.P., Halmanov, H.: On the theory of fatigue crack growth. Engineer-
     ing Fracture Mechanics. 4, 219–230 (1972). https://doi.org/10.1016/0013-7944(72)900
     38-0.

295. Ulbin, M., Glodež, S., Vesenjak, M., Duarte, I., Podgornik, B., Ren, Z., Kramberger, J.: Low cycle fatigue behaviour of closed-cell aluminium foam. Mechanics of Materials. 133, 165–173 (2019). https://doi.org/10.1016/j.mechmat.2019.03.014.

296. Sukumar, N., Dolbow, J.E., Moës, N.: Extended finite element method in computational fracture mechanics: A retrospective examination. International Journal of Fracture. 196, 189–206 (2015). https://doi.org/10.1007/s10704-015-0064-8.

297. Ashcroft, I.A., Mubashar, A.: Numerical approach: Finite element analysis. In: da Silva, L.F.M., Öchsner, A., and Adams, R.D. (eds.) Handbook of Adhesion Technology. pp. 629–660. Springer Berlin Heidelberg, Berlin, Heidelberg (2011). https://doi.org/10.1007/978-3-642-01169-6_25.

298. Mzabi, S., Berghezan, D., Roux, S., Hild, F., Creton, C.: A critical local energy release rate criterion for fatigue fracture of elastomers. Journal of Polymer Science Part B: Polymer Physics. 49, 1518–1524 (2011). https://doi.org/10.1002/polb.22338.

299. Silitonga, S., Soetens, F., Snijder, H.H., Maljaars, J.: Fatigue life estimation of metal structures based on damage modeling, https://research.tue.nl/en/publications/fatigue-life-estimation-of-metal-structures-based-on-damage-model, (2017).

300. Ambriz, R.R., Mesmacque, G., Ruiz, A., Amrouche, A., López, V.H., Benseddiq, N.: Fatigue crack growth under a constant amplitude loading of Al-6061-T6 welds obtained by modified indirect electric arc technique. Science and Technology of Welding and Joining. 15, 514–521 (2010). https://doi.org/10.1179/136217110X12785889549589.

301. Adalsteinsson, D., Sethian, J.A.: A fast level set method for propagating interfaces. Journal of Computational Physics. 118, 269–277 (1995). https://doi.org/10.1006/jcph.1995.1098.

302. Dunham, W.: Journey through genius: The great theorems of mathematics. Penguin Books, New York, NY (1991).

303. Sabelkin, V., Mall, S., Avram, J.B.: Fatigue crack growth analysis of stiffened cracked panel repaired with bonded composite patch. Engineering Fracture Mechanics. 73, 1553–1567 (2006). https://doi.org/10.1016/j.engfracmech.2006.01.029.

304. Awd, M., Labanie, M.F., Moehring, K., Fatemi, A., Walther, F.: Towards deterministic computation of internal stresses in additively manufactured materials under fatigue loading: Part I. Materials. 13, 2318 (2020). https://doi.org/10.3390/ma13102318.

305. Smith, M.: Abaqus/Standard User's Manual, Version 6.9. Dassault Systèmes Simulia Corp, United States (2009).

306. Gracie, R., Belytschko, T.: Concurrently coupled atomistic and XFEM models for dislocations and cracks. International Journal for Numerical Methods in Engineering. 78, 354–378 (2009). https://doi.org/10.1002/nme.2488.

307. Yi, G., Sui, Y., Du, J.: Application of Python-based Abaqus preprocess and postprocess technique in analysis of gearbox vibration and noise reduction. Frontiers of Mechanical Engineering. 6, 229 (2011). https://doi.org/10.1007/s11465-011-0128-z.

308. Farahani, B.V., Tavares, P.J., Belinha, J., Moreira, P.M.G.P.: A fracture mechanics study of a compact tension specimen: Digital image correlation, finite element and meshless methods. Procedia Structural Integrity. 5, 920–927 (2017). https://doi.org/10.1016/j.prostr.2017.07.113.

309. Raabe, D. ed: Continuum scale simulation of engineering materials: Fundamentals, microstructures, process applications. Wiley-VCH, Weinheim (2004).

310. Dunne, F., Petrinic, N.: Introduction to computational plasticity. Oxford University Press, Oxford; New York (2005).

311. Trapp, M., Öchsner, A.: Computational plasticity for finite elements. Springer Berlin Heidelberg, New York, NY (2018).

312. Bari, S., Hassan, T.: Kinematic hardening rules in uncoupled modeling for multiaxial ratcheting simulation. International Journal of Plasticity. 17, 885–905 (2001). https://doi.org/10.1016/S0749-6419(00)00031-0.

313. Alejano, L.R., Bobet, A.: Drucker–Prager criterion. Rock Mechanics and Rock Engineering. 45, 995–999 (2012). https://doi.org/10.1007/s00603-012-0278-2.

314. Chaboche, J.L.: Time-independent constitutive theories for cyclic plasticity. International Journal of Plasticity. 2, 149–188 (1986). https://doi.org/10.1016/0749-6419(86)90010-0.

315. Badnava, H., Pezeshki, S.M., Fallah Nejad, Kh., Farhoudi, H.R.: Determination of combined hardening material parameters under strain controlled cyclic loading by using the genetic algorithm method. Journal of Mechanical Science and Technology. 26, 3067–3072 (2012). https://doi.org/10.1007/s12206-012-0837-1.

316. Kaleva, O., Orelma, H., Petukhov, D.: Parameter estimation of a high-cycle fatigue model combining the Ottosen-Stenström-Ristinmaa approach and Lemaitre-Chaboche damage rule. International Journal of Fatigue. 147, 106153 (2021). https://doi.org/10.1016/j.ijfatigue.2021.106153.

317. Lemaitre, J., Chaboche, J.-L., Shrivastava, B.: Mechanics of solid materials. Cambridge University Press, Cambridge (2002).

318. Mahmoudi, A.H., Pezeshki-Najafabadi, S.M., Badnava, H.: Parameter determination of Chaboche kinematic hardening model using a multi objective genetic algorithm. Computational Materials Science. 50, 1114–1122 (2011). https://doi.org/10.1016/j.commatsci.2010.11.010.

319. Wang, H., Yan, Y., Wan, M., Wu, X.: Experimental investigation and constitutive modeling for the hardening behavior of 5754O aluminum alloy sheet under two-stage loading. International Journal of Solids and Structures. 49, 3693–3710 (2012). https://doi.org/10.1016/j.ijsolstr.2012.08.007.

320. Gozzi, J., Olsson, A.: Extra high strength steel plasticity-experimental work and constitutive modelling. In: Fourth International Conference on Advances in Steel Structures. pp. 1571–1576. Elsevier (2005). https://doi.org/10.1016/B978-008044637-0/50234-1.

321. Zhao, K.M., Lee, J.K.: Finite element analysis of the three-point bending of sheet metals. Journal of Materials Processing Technology. 122, 6–11 (2002). https://doi.org/10.1016/S0924-0136(01)01064-0.

322. Fathi, D.M., Okail, H.O., Mahdi, H.A., Abdelrahman, A.A.: Cyclic load behavior of precast self-centering hammer head bridge piers. HBRC Journal. 16, 113–141 (2020). https://doi.org/10.1080/16874048.2020.1789385.

323. Novak, J.S., Benasciutti, D., De Bona, F., Stanojević, A., De Luca, A., Raffaglio, Y.: Estimation of material parameters in nonlinear hardening plasticity models and strain life curves for CuAg alloy. IOP Conference Series: Materials Science and Engineering. 119, 012020 (2016). https://doi.org/10.1088/1757-899X/119/1/012020.

324. Biswas, A., Prasad, M.R.G., Vajragupta, N., ul Hassan, H., Brenne, F., Niendorf, T., Hartmaier, A.: Influence of microstructural features on the strain hardening behavior of additively manufactured metallic components. Advanced Engineering Materials. 21, 1900275 (2019). https://doi.org/10.1002/adem.201900275.

325. Zhao, Z., To, S.: An investigation of resolved shear stress on activation of slip systems during ultraprecision rotary cutting of local anisotropic Ti-6Al-4V alloy: Models and experiments. International Journal of Machine Tools and Manufacture. 134, 69–78 (2018). https://doi.org/10.1016/j.ijmachtools.2018.07.001.

326. Agius, D., Kourousis, K.I., Wallbrink, C.: Elastoplastic response of as-built SLM and wrought Ti-6Al-4V under symmetric and asymmetric strain-controlled cyclic loading. Rapid Prototyping Journal. 24, 1409–1420 (2018). https://doi.org/10.1108/RPJ-05-2017-0105.

327. Muth, A., John, R., Pilchak, A., Kalidindi, S.R., McDowell, D.L.: Analysis of fatigue indicator parameters for Ti-6Al-4V microstructures using extreme value statistics in the HCF regime. International Journal of Fatigue. 145, 106096 (2021). https://doi.org/10.1016/j.ijfatigue.2020.106096.

328. Fatemi, A., Kurath, P.: Multiaxial fatigue life predictions under the influence of mean-stresses. Journal of Engineering Materials and Technology. 110, 380–388 (1988). https://doi.org/10.1115/1.3226066.

329. Fatemi, A., Molaei, R., Phan, N.: Multiaxial fatigue of additive manufactured metals: Performance, analysis, and applications. International Journal of Fatigue. 134, 105479 (2020). https://doi.org/10.1016/j.ijfatigue.2020.105479.

330. Roters, F., Eisenlohr, P., Hantcherli, L., Tjahjanto, D.D., Bieler, T.R., Raabe, D.: Overview of constitutive laws, kinematics, homogenization and multiscale methods in crystal plasticity finite-element modeling: Theory, experiments, applications. Acta Materialia. 58, 1152–1211 (2010). https://doi.org/10.1016/j.actamat.2009.10.058.

331. Budynas, R.G., Nisbett, J.K., Shigley, J.E.: Shigley's mechanical engineering design. McGraw-Hill Education, New York, NY (2015).

332. Karolczuk, A., Papuga, J., Palin-Luc, T.: Progress in fatigue life calculation by implementing life-dependent material parameters in multiaxial fatigue criteria. International Journal of Fatigue. 134, 105509 (2020). https://doi.org/10.1016/j.ijfatigue.2020.105509.

333. Stopka, K.S., McDowell, D.L.: Microstructure-sensitive computational multiaxial fatigue of Al 7075-T6 and duplex Ti-6Al-4V. International Journal of Fatigue. 133, 105460 (2020). https://doi.org/10.1016/j.ijfatigue.2019.105460.

334. Chen, J.H., Cao, R.: Micromechanism of cleavage fracture of metals: A comprehensive microphysical model for cleavage cracking in metals. Elsevier, Butterworth-Heinemann, Amsterdam; Boston (2015).

335. Lütjering, G., Williams, J.C.: Titanium. Springer Berlin Heidelberg, Berlin, Heidelberg (2007). https://doi.org/10.1007/978-3-540-73036-1.

336. Ou, C.-Y., Voothaluru, R., Liu, C.R.: Fatigue crack initiation of metals fabricated by additive manufacturing — A crystal plasticity energy-based approach to IN718 life prediction. Crystals. 10, 905 (2020). https://doi.org/10.3390/cryst10100905.

337. Mlikota, M., Schmauder, S.: A newly discovered relation between the critical resolved shear stress and the fatigue endurance limit for metallic materials. Metals. 10, 803 (2020). https://doi.org/10.3390/met10060803.

338. Nihei, M., Heuler, P., Boller, C., Seeger, T.: Evaluation of mean stress effect on fatigue life by use of damage parameters. International Journal of Fatigue. 8, 119–126 (1986). https://doi.org/10.1016/0142-1123(86)90002-2.

339. Esmaeili, F., Zehsaz, M., Chakherlou, T.N., Barzegar, S.: Fatigue life estimation of double lap simple bolted and hybrid (bolted/bonded) joints using several multiaxial fatigue criteria. Materials & Design. 67, 583–595 (2015). https://doi.org/10.1016/j.mat des.2014.11.003.

340. Troshchenko, V.T.: Fatigue of metals under nonuniform stressed state. Part 2: Methods of the analysis of research results. Strength of Materials. 42, 241–257 (2010). https://doi.org/10.1007/s11223-010-9213-5.

341. McDiarmid, D.L.: A general criterion for high cycle multiaxial fatigue failure. Fatigue & Fracture of Engineering Materials and Structures. 14, 429–453 (1991). https://doi.org/10.1111/j.1460-2695.1991.tb00673.x.

342. Chu, C.-C., Conle, F., Bonnen, J.: Multiaxial stress-strain modeling and fatigue life prediction of SAE axle shafts. In: McDowell, D. and Ellis, J. (eds.) Advances in Multiaxial Fatigue. pp. 37–37–18. ASTM International, 100 Barr Harbor Drive, PO Box C700, West Conshohocken, PA 19428–2959 (1993). https://doi.org/10.1520/STP24794S.

343. Broggiato, G.B., Campana, F., Cortese, L.: The Chaboche nonlinear kinematic hardening model: Calibration methodology and validation. Meccanica. 43, 115–124 (2008). https://doi.org/10.1007/s11012-008-9115-9.

344. Yun, G.J., Shang, S.: A self-optimizing inverse analysis method for estimation of cyclic elasto-plasticity model parameters. International Journal of Plasticity. 27, 576–595 (2011). https://doi.org/10.1016/j.ijplas.2010.08.003.

345. Socie, D., Waill, L., Dittmer, D.: Biaxial fatigue of Inconel 718 including mean stress effects. In: Miller, K. and Brown, M. (eds.) Multiaxial Fatigue. pp. 463–463–19. ASTM International, 100 Barr Harbor Drive, PO Box C700, West Conshohocken, PA 19428–2959 (1985). https://doi.org/10.1520/STP36238S.

346. Fatemi, A., Socie, D.F.: A critical plane approach to multiaxial fatigue damage including out-of-phase loading. Fatigue & Fracture of Engineering Materials and Structures. 11, 149–165 (1988). https://doi.org/10.1111/j.1460-2695.1988.tb01169.x.

347. Chen, X., Gao, Q., Sun, X.-F.: Low-cycle fatigue under non-proportional loading. Fatigue & Fracture of Engineering Materials and Structures. 19, 839–854 (1996). https://doi.org/10.1111/j.1460-2695.1996.tb01020.x.

348. Gates, N.R., Fatemi, A.: On the consideration of normal and shear stress interaction in multiaxial fatigue damage analysis. International Journal of Fatigue. 100, 322–336 (2017). https://doi.org/10.1016/j.ijfatigue.2017.03.042.

349. Karolczuk, A., Skibicki, D., Pejkowski, Ł.: Evaluation of the Fatemi-Socie damage parameter for the fatigue life calculation with application of the Chaboche plasticity model. Fatigue & Fracture of Engineering Materials & Structures. 42, 197–208 (2019). https://doi.org/10.1111/ffe.12895.

350. Marines, I.: An understanding of very high cycle fatigue of metals. International Journal of Fatigue. 25, 1101–1107 (2003). https://doi.org/10.1016/S0142-1123(03)001 47-6.

351. Przybyla, C.P., Musinski, W.D., Castelluccio, G.M., McDowell, D.L.: Microstructure-sensitive HCF and VHCF simulations. International Journal of Fatigue. 57, 9–27 (2013). https://doi.org/10.1016/j.ijfatigue.2012.09.014.

352. Liu, Y.B., Li, Y.D., Li, S.X., Yang, Z.G., Chen, S.M., Hui, W.J., Weng, Y.Q.: Prediction of the S–N curves of high-strength steels in the very high cycle fatigue regime. International Journal of Fatigue. 32, 1351–1357 (2010). https://doi.org/10.1016/j.ijfatigue.2010.02.006.

353. Zimmermann, M.: Diversity of damage evolution during cyclic loading at very high numbers of cycles. International Materials Reviews. 57, 73–91 (2012). https://doi.org/10.1179/1743280411Y.0000000005.

354. Przybyla, C.P., McDowell, D.L.: Microstructure-sensitive extreme value probabilities for high cycle fatigue of Ni-base superalloy IN100. International Journal of Plasticity. 26, 372–394 (2010). https://doi.org/10.1016/j.ijplas.2009.08.001.

355. Przybyla, C., Prasannavenkatesan, R., Salajegheh, N., McDowell, D.L.: Microstructure-sensitive modeling of high cycle fatigue. International Journal of Fatigue. 32, 512–525 (2010). https://doi.org/10.1016/j.ijfatigue.2009.03.021.

356. Maktouf, W., Ammar, K., Ben Naceur, I., Saï, K.: Multiaxial high-cycle fatigue criteria and life prediction: Application to gas turbine blade. International Journal of Fatigue. 92, 25–35 (2016). https://doi.org/10.1016/j.ijfatigue.2016.06.024.

357. Stephens, R.I., Fuchs, H.O. eds: Metal fatigue in engineering. Wiley, New York (2001).

358. Bennett, V., Mcdowell, D.: Polycrystal orientation distribution effects on microslip in high cycle fatigue. International Journal of Fatigue. 25, 27–39 (2003). https://doi.org/10.1016/S0142-1123(02)00057-9.

359. Liu, Y., Mahadevan, S.: Multiaxial high-cycle fatigue criterion and life prediction for metals. International Journal of Fatigue. 27, 790–800 (2005). https://doi.org/10.1016/j.ijfatigue.2005.01.003.

360. Salajegheh, N., McDowell, D.L.: Microstructure-sensitive weighted probability approach for modeling surface to bulk transition of high cycle fatigue failures dominated by primary inclusions. International Journal of Fatigue. 59, 188–199 (2014). https://doi.org/10.1016/j.ijfatigue.2013.08.025.

361. Przybyla, C.P., McDowell, D.L.: Simulated microstructure-sensitive extreme value probabilities for high cycle fatigue of duplex Ti–6Al–4V. International Journal of Plasticity. 27, 1871–1895 (2011). https://doi.org/10.1016/j.ijplas.2011.01.006.

362. Przybyla, C.P., McDowell, D.L.: Microstructure-sensitive extreme-value probabilities of high-cycle fatigue for surface vs. subsurface crack formation in duplex Ti–6Al–4V. Acta Materialia. 60, 293–305 (2012). https://doi.org/10.1016/j.actamat.2011.09.031.

363. Burago, N., Nikitin, I.: Multiaxial fatigue criteria and durability of titanium compressor disks in low- and very-high-cycle fatigue modes. In: Neittaanmäki, P., Repin, S., and Tuovinen, T. (eds.) Mathematical Modeling and Optimization of Complex Structures. pp. 117–130. Springer International Publishing, Cham (2016). https://doi.org/10.1007/978-3-319-23564-6_8.

364. Castelluccio, G.M., McDowell, D.L.: Effect of annealing twins on crack initiation under high cycle fatigue conditions. Journal of Materials Science. 48, 2376–2387 (2013). https://doi.org/10.1007/s10853-012-7021-y.

365. Arora, P., Gupta, S.K., Samal, M.K., Chattopadhyay, J.: Multiaxial fatigue tests under variable strain paths and asynchronous loading and assessment of fatigue life using critical plane models. International Journal of Fatigue. 145, 106049 (2021). https://doi.org/10.1016/j.ijfatigue.2020.106049.

366. Yu, Q., Zhang, J., Jiang, Y., Li, Q.: Multiaxial fatigue of extruded AZ61A magnesium alloy. International Journal of Fatigue. 33, 437–447 (2011). https://doi.org/10.1016/j.ijfatigue.2010.09.020.

367. Johnson, C.A.: MITI and the Japanese miracle: The growth of industrial policy, 1925 – 1975. Stanford University Press, Stanford, California (2007).

368. Walpole, R.E., Myers, R.H., Myers, S.L., Ye, K.: Probability & statistics for engineers & scientists: MyStatLab update. (2017).

369. Jacod, J., Protter, P.E.: Probability essentials. Springer, Berlin; New York (2004).

370. Spanos, P.D., Wu, Y.-T.: Probabilistic structural mechanics: Advances in structural reliability methods. Springer Berlin Heidelberg, Berlin, Heidelberg (1994).

371. Hoff, P.D.: A first course in Bayesian statistical methods. Springer, London; New York (2009).

372. Castillo, E., Fernández-Canteli, A.: A unified statistical methodology for modeling fatigue damage. Springer, Dordrecht (2009).

373. Sobczyk, K.: Stochastic approach to fatigue: Experiments, modelling and reliability estimation. Springer Wien, Vienna (2014).

374. Castillo, E., Galambos, J., Sarabia, J.M.: The selection of the domain of attraction of an extreme value distribution from a set of data. In: Hüsler, J. and Reiss, R.-D. (eds.) Extreme Value Theory. pp. 181–190. Springer New York, New York, NY (1989). https://doi.org/10.1007/978-1-4612-3634-4_16.

375. Wang, B., Xie, L., Song, J., Zhao, B., Li, C., Zhao, Z.: Curved fatigue crack growth prediction under variable amplitude loading by artificial neural network. International Journal of Fatigue. 142, 105886 (2021). https://doi.org/10.1016/j.ijfatigue.2020. 105886.

376. Boyer, R.R.: Attributes, characteristics, and applications of titanium and its alloys. Journal of The Minerals, Metals & Materials Society (TMS). 62, 21–24 (2010). https://doi.org/10.1007/s11837-010-0071-1.

377. Castillo, E., Hadi, A.S.: A method for estimating parameters and quantiles of distributions of continuous random variables. Computational Statistics & Data Analysis. 20, 421–439 (1995). https://doi.org/10.1016/0167-9473(94)00049-O.

378. Wormsen, A., Sjödin, B., Härkegård, G., Fjeldstad, A.: Non-local stress approach for fatigue assessment based on weakest-link theory and statistics of extremes. Fatigue & Fracture of Engineering Materials and Structures. 30, 1214–1227 (2007). https://doi.org/10.1111/j.1460-2695.2007.01190.x.

379. Pyttel, B., Schwerdt, D., Berger, C.: Very high cycle fatigue—Is there a fatigue limit? International Journal of Fatigue. 33, 49–58 (2011). https://doi.org/10.1016/j.ijfatigue. 2010.05.009.

380. Maltamo, M., Kangas, A., Uuttera, J., Torniainen, T., Saramäki, J.: Comparison of percentile based prediction methods and the Weibull distribution in describing the diameter distribution of heterogeneous Scots pine stands. Forest Ecology and Management. 133, 263–274 (2000). https://doi.org/10.1016/S0378-1127(99)00239-X.

381. Coleman, B.D.: Statistics and time dependence of mechanical breakdown in fibers. Journal of Applied Physics. 29, 968–983 (1958). https://doi.org/10.1063/1.1723343.

382. Jenkinson, A.F.: The frequency distribution of the annual maximum (or minimum) values of meteorological elements. Quarterly Journal of the Royal Meteorological Society. 81, 158–171 (1955). https://doi.org/10.1002/qj.49708134804.

383. Hosking, J.R.M., Wallis, J.R., Wood, E.F.: Estimation of the generalized extreme-value distribution by the method of probability-weighted moments. Technometrics. 27, 251–261 (1985). https://doi.org/10.1080/00401706.1985.10488049.

384. Wasserman, L.: All of statistics: A concise course in statistical inference. Springer, New York Berlin Heidelberg (2010).

385. Robert, C.: The Bayesian choice: From decision-theoretic foundations to computational implementation. Springer, New York (2007).

386. Cowles, M.K.: Applied Bayesian statistics: With R and OpenBUGS examples. Springer, New York (2013).

387. Durodola, J.F., Ramachandra, S., Gerguri, S., Fellows, N.A.: Artificial neural network for random fatigue loading analysis including the effect of mean stress. International Journal of Fatigue. 111, 321–332 (2018). https://doi.org/10.1016/j.ijfatigue.2018.02.007.

388. Zhang, M., Sun, C.-N., Zhang, X., Goh, P.C., Wei, J., Hardacre, D., Li, H.: High cycle fatigue life prediction of laser additive manufactured stainless steel: A machine learning approach. International Journal of Fatigue. 128, 105194 (2019). https://doi.org/10.1016/j.ijfatigue.2019.105194.

389. Nowell, D., Nowell, P.W.: A machine learning approach to the prediction of fretting fatigue life. Tribology International. 141, 105913 (2020). https://doi.org/10.1016/j.triboint.2019.105913.

390. Hu, D., Su, X., Liu, X., Mao, J., Shan, X., Wang, R.: Bayesian-based probabilistic fatigue crack growth evaluation combined with machine-learning-assisted GPR. Engineering Fracture Mechanics. 229, 106933 (2020). https://doi.org/10.1016/j.engfracmech.2020.106933.

391. Chen, J., Liu, Y.: Probabilistic physics-guided machine learning for fatigue data analysis. Expert Systems with Applications. 168, 114316 (2021). https://doi.org/10.1016/j.eswa.2020.114316.

392. Zhan, Z., Li, H.: A novel approach based on the elastoplastic fatigue damage and machine learning models for life prediction of aerospace alloy parts fabricated by additive manufacturing. International Journal of Fatigue. 145, 106089 (2021). https://doi.org/10.1016/j.ijfatigue.2020.106089.

393. Shiraiwa, T., Briffod, F., Miyazawa, Y., Enoki, M.: Fatigue performance prediction of structural materials by multi-scale modeling and machine learning. In: Mason, P., Fisher, C.R., Glamm, R., Manuel, M.V., Schmitz, G.J., Singh, A.K., and Strachan, A. (eds.) Proceedings of the 4th World Congress on Integrated Computational Materials Engineering (ICME 2017). pp. 317–326. Springer International Publishing, Cham (2017). https://doi.org/10.1007/978-3-319-57864-4_29.

394. Cui, L., Jiang, F., Peng, R.L., Mousavian, R.T., Yang, Z., Moverare, J.: Dependence of microstructures on fatigue performance of polycrystals: A comparative study of conventional and additively manufactured 316L stainless steel. International Journal of Plasticity. 149, 103172 (2022). https://doi.org/10.1016/j.ijplas.2021.103172.

395. Reimann, D., Nidadavolu, K., ul Hassan, H., Vajragupta, N., Glasmachers, T., Junker, P., Hartmaier, A.: Modeling macroscopic material behavior with machine learning algorithms trained by micromechanical simulations. Frontiers in Materials. 6, 181 (2019). https://doi.org/10.3389/fmats.2019.00181.

396. Zhan, Z., Li, H.: Machine learning based fatigue life prediction with effects of additive manufacturing process parameters for printed SS 316L. International Journal of Fatigue. 142, 105941 (2021). https://doi.org/10.1016/j.ijfatigue.2020.105941.

397. Jang, D.-W., Lee, S., Park, J.-W., Baek, D.-C.: Failure detection technique under random fatigue loading by machine learning and dual sensing on symmetric structure. International Journal of Fatigue. 114, 57–64 (2018). https://doi.org/10.1016/j.ijfatigue.2018.05.004.

398. Keprate, A., Ratnayake, R.M.C.: Artificial intelligence based approach for predicting fatigue strength using composition and process parameters. In: Volume 3: Materials Technology. p. V003T03A017. American Society of Mechanical Engineers, Virtual, Online (2020). https://doi.org/10.1115/OMAE2020-18675.

399. Shipley, H., McDonnell, D., Culleton, M., Coull, R., Lupoi, R., O'Donnell, G., Trimble, D.: Optimisation of process parameters to address fundamental challenges during selective laser melting of Ti-6Al-4V: A review. International Journal of Machine Tools and Manufacture. 128, 1–20 (2018). https://doi.org/10.1016/j.ijmachtools.2018.01.003.

400. Choy, S.Y., Sun, C.-N., Leong, K.F., Wei, J.: Compressive properties of functionally graded lattice structures manufactured by selective laser melting. Materials & Design. 131, 112–120 (2017). https://doi.org/10.1016/j.matdes.2017.06.006.

401. Thijs, L., Verhaeghe, F., Craeghs, T., Humbeeck, J.V., Kruth, J.-P.: A study of the microstructural evolution during selective laser melting of Ti–6Al–4V. Acta Materialia. 58, 3303–3312 (2010). https://doi.org/10.1016/j.actamat.2010.02.004.

402. Mishurova, T., Artzt, K., Haubrich, J., Requena, G., Bruno, G.: New aspects about the search for the most relevant parameters optimizing SLM materials. Additive Manufacturing. 25, 325–334 (2019). https://doi.org/10.1016/j.addma.2018.11.023.

403. Gong, H., Rafi, K., Gu, H., Starr, T., Stucker, B.: Analysis of defect generation in Ti–6Al–4V parts made using powder bed fusion additive manufacturing processes. Additive Manufacturing. 1–4, 87–98 (2014). https://doi.org/10.1016/j.addma.2014.08.002.

404. Ma, X., He, X., Tu, Z.C.: Prediction of fatigue–crack growth with neural network-based increment learning scheme. Engineering Fracture Mechanics. 241, 107402 (2021). https://doi.org/10.1016/j.engfracmech.2020.107402.

405. Haque, M.E., Sudhakar, K.V.: ANN based prediction model for fatigue crack growth in DP steel. Fatigue & Fracture of Engineering Materials & Structures. 24, 63–68 (2001). https://doi.org/10.1046/j.1460-2695.2001.00361.x.

406. Younis, H.B., Kamal, K., Sheikh, M.F., Hamza, A., Zafar, T.: Prediction of fatigue crack growth rate in aircraft aluminum alloys using radial basis function neural network. In: 2018 Tenth International Conference on Advanced Computational Intelligence (ICACI). pp. 825–830. IEEE, Xiamen (2018). https://doi.org/10.1109/ICACI.2018.8377568.

407. Herzog, D., Seyda, V., Wycisk, E., Emmelmann, C.: Additive manufacturing of metals. Acta Materialia. 117, 371–392 (2016). https://doi.org/10.1016/j.actamat.2016.07.019.

408. Spierings, A.B., Dawson, K., Voegtlin, M., Palm, F., Uggowitzer, P.J.: Microstructure and mechanical properties of as-processed scandium-modified aluminium using selective laser melting. CIRP Annals. 65, 213–216 (2016). https://doi.org/10.1016/j.cirp.2016.04.057.

409. El-tawel, Gh.S., Helmy, A.K.: An edge detection scheme based on least squares support vector machine in a contourlet HMT domain. Applied Soft Computing. 26, 418–427 (2015). https://doi.org/10.1016/j.asoc.2014.10.025.

Related Student Works[1,2]

1. Tung, C.: Development of cyclic plasticity models for selective laser melted lightweight alloys. Master Thesis. TU Dortmund University (2019).
2. Merghany, M.: Generation of two- and three-dimensional representative volume elements for selective laser melted Ti-6Al-4V and AlSi10Mg. Project Work. TU Dortmund University (2021).
3. Zhang, Z.: Numerical analysis of crack growth using XFEM and contour integral methods. Project Work. TU Dortmund University (2021).
4. Taha, A.: Computation of probabilistic percentiles for *S-N* curves of additively manufactured metallic parts. Project Work. TU Dortmund University (2021).
5. Sengün, B.: Determination of the crack propagation rate in additively manufactured Al and Ti alloys through XFEM simulations. Project Work. TU Dortmund University (2022).
6. Sengün, B.: Investigation of influence of testing conditions and elastic properties on fatigue crack propagation rate using XFEM method. Master Thesis. TU Dortmund University (2022).

Related Publications

1. **Awd, M.;** Münstermann, S.; Walther, F.: Effect of microstructural heterogeneity on fatigue strength predicted by reinforcement machine learning. Fatigue and Fracture of Engineering Materials and Structures. 45, 3267–3287 (2022). https://doi.org/10.1111/ffe.13816
2. **Awd, M.;** Walther, F.: Numerical investigation of the influence of fatigue testing frequency on the fracture and crack propagation rate of additive-manufactured AlSi10Mg and Ti-6Al-4V alloys. Solids. 3, 430–446 (2022). https://doi.org/10.3390/solids3030030
3. **Awd, M.;** Walther, F.: Influence of fatigue testing frequency on fracture and crack propagation rate in additive manufactured metals using the extended finite element method. Proceedings of the 5th Iberian Conference on Structural Integrity, Coimbra, Portugal (2022) 1–6.
4. **Awd, M.;** Siddique, S.; Walther, F.: Microstructure-based assessment of damage mechanisms of selective laser melted Al-Si alloys under very high-cycle fatigue loading. VHCF8, Proceedings of the 8th International Conference on Very High Cycle Fatigue (2021) Online, Sapporo, Japan, 1–7.
5. **Awd, M.;** Walther, F.; Siddique, S.; Fatemi, A.: Microstructure and fatigue damage evolution in additive-manufactured metals using enhanced measurement techniques and modeling approaches. TMS 2021 150th Annual Meeting & Exhibition Supplemental Proceedings, Springer International Publishing (2021) 753-762. https://doi.org/10.1007/978-3-030-65261-6_68

---

[1] In the subject area of the dissertation, the author supervised the following student theses, among others not related to the topic of this dissertation.

[2] I would like to thank the students for their contributions.

6. Stern, F.; Lüneburg, B.; Rollmann, J.; Pape, F.; Pantke, K.; **Awd, M.;** Tenkamp, J.; Walther, F.: Evaluation of the influence of non-metallic inclusions on the high and very high cycle fatigue life of inductive hardened bearing steel in multi-megawatt wind turbines. Proceedings of Bearing World 2020, International Bearing Conference (2020) 67–70.

7. Tenkamp, J.; **Awd, M.;** Siddique, S.; Starke, P.; Walther, F.: Fracture–mechanical assessment of the effect of defects on the fatigue lifetime and limit in cast and additively manufactured aluminum–silicon alloys from HCF to VHCF regime. Metals 10(7), 943 (2020) 1–18. https://doi.org/10.3390/met10070943

8. **Awd, M.;** Labanie, F.; Möhring, K.; Fatemi, A.; Walther, F.: Towards deterministic computation of internal stresses in additively manufactured materials under fatigue loading: Part I. Materials 13(10), 2318 (2020) 1–15. https://doi.org/10.3390/ma13102318. **Invited Contribution**

9. **Awd, M.;** Johannsen, J.; Chan, T.; Merghany, M.; Emmelmann, C.; Walther, F.: Improvement of fatigue strength in lightweight selective laser melted alloys by in situ and ex-situ composition and heat treatment. TMS 2020 149th Annual Meeting and Exhibition Supplemental Proceedings, Springer International Publishing Cham (2020) 115-126. https://doi.org/10.1007/978-3-030-36296-6_11

10. **Awd, M.;** Siddique, S.; Walther, F.: Microstructural damage and fracture mechanisms of selective laser melted Al-Si alloys under fatigue loading. Theoretical and Applied Fracture Mechanics 106 (2020) 102483. Invited Contribution. https://doi.org/10.1016/j.tafmec.2020.102483

11. Kotzem, D.; Beermann, L.; **Awd, M.;** Walther, F.: Mechanical and microstructural characterization of arc-welded Inconel 625 alloy. Materials 12(22), 3690 (2019) 1–10. https://doi.org/10.3390/ma12223690

12. **Awd, M.;** Merghany, M.; Siddique, S.; Walther, F.: Simulation-based investigation of statistical fatigue strength of selective laser melted lightweight alloys. Proceedings of the 2nd International Conference on Simulation for Additive Manufacturing (2019) 262–273. ISBN: 978-84-949194-8-0.

13. **Awd, M.;** Merghany, M.; Johannsen, J.; Emmelmann, C.; Walther, F.: Microstructural damage and fracture mechanisms of selective laser melted AlSi10Mg alloy under cyclic loading. Proceedings of the 1st European Conference on Structural Integrity of Additively Manufactured Materials (2019) 1–6.

14. Biswal, R.; Zhang, X.; Shamir, M.; Al Mamun, A.; **Awd, M.;** Walther, F.; Syed, A.: Interrupted fatigue testing with periodic tomography to monitor porosity defects in wire + arc additive manufactured Ti-6Al-4V. Additive Manufacturing 8 (2019) 517-527. https://doi.org/10.1016/j.addma.2019.04.026

15. **Awd, M.;** Siddique, S.; Johannsen, J.; Emmelmann, C.; Walther, F.: Very high-cycle fatigue properties and microstructural damage mechanisms of selective laser melted AlSi10Mg alloy. International Journal of Fatigue 124 (2019) 55-69. https://doi.org/10.1016/j.ijfatigue.2019.02.040

16. Syed, A.; **Awd, M.;** Walther, F.; Zhang, X.: Microstructure and mechanical properties of as-built and heat-treated electron beam melted Ti-6Al-4V. Materials Science and Technology 35,6 (2019) 653-660. https://doi.org/10.1080/02670836.2019.1580434

17. Biswal, R.; Zhang, X.; Syed, A.; **Awd, M.;** Ding, J.; Walther, F.; Williams, S.: Criticality of porosity defects on the fatigue performance of wire + arc additive manufactured titanium alloy. International Journal of Fatigue 122 (2019) 208-217. https://doi.org/10.1016/j.ijfatigue.2019.01.017

18. **Awd, M.;** Siddique, S.; Johannsen, J.; Wiegold, T.; Klinge, S.; Emmelmann, C.; Walther, F.: Quality assurance of additively manufactured alloys for aerospace industry by non-destructive testing and numerical modeling. Proceedings of the 10th International Conference on Non-destructive Testing in Aerospace (2018) 1–10.

19. Siddique, S.; **Awd, M.;** Wiegold, T.; Klinge, S.; Walther, F.: Simulation of cyclic deformation behavior of selective laser melted and hybrid-manufactured aluminum alloys using the phase-field method. Applied Sciences 8(10), 1948 (2018) 1–18. https://doi.org/10.3390/app8101948

20. **Awd, M.;** Stern, F.; Kampmann, A.; Kotzem, D.; Tenkamp, J.; Walther, F.: Microstructural characterization of the anisotropy and cyclic deformation behavior of selective laser melted AlSi10Mg structures. Metals 8(10), 825 (2018) 1–14. https://doi.org/10.3390/met8100825

21. **Awd, M.;** Siddique, S.; Hajavifard, R.; Walther, F.: Comparative study of defect-based and plastic damage-based approaches for fatigue lifetime calculation of selective laser-melted AlSi12. FFW 2018, Proceedings of the 7th International Conference on Fracture Fatigue and Wear (2018) 297–313. https://doi.org/10.1007/978-981-13-0411-8_27. **Best Paper Award.**

22. **Awd, M.;** Johannsen, J.; Siddique, S.; Emmelmann, C.; Walther, F.: Qualification of selective laser-melted Al alloys against fatigue damage by means of measurement and modeling techniques. Fatigue 2018, Matec Web of Conferences 165, 02001 (2018) 1–7. https://doi.org/10.1051/matecconf/201816502001

23. Frömel, F.; **Awd, M.;** Tenkamp, J.; Walther, F.; Boes, J.; Geenen, K.; Röttger, A.; Theisen, W.: Untersuchung des Einflusses einer HIP-Nachbehandlung auf den selektiv lasergeschmolzenen (SLM) austenitischen Stahl AISI 316L. Werkstoffe und Additive Fertigung—Tagungsband, Hrsg: P. Hoyer, C. Leyens, T. Niendorf, V. Ploshikhin, V. Schulze, G. Witt, ISBN 978–3–88355–418–1 (2018) 190–199.

24. **Awd, M.;** Tenkamp, J.; Hirtler, M.; Siddique, S.; Bambach, B.; Walther, F.: Comparison of microstructure and mechanical properties of Scalmalloy® produced by selective laser melting and laser metal deposition. Materials 11, 1 (2018) 1–17. https://doi.org/10.3390/ma11010017

25. **Awd, M.;** Siddique, S.; Tenkamp, J.; Walther, F.: Freeform characterization of fatigue strength of additively manufactured lightweight alloys through FEM and Monte-Carlo modeling. 2. Tagung des DVM-AK Additiv gefertigte Bauteile und Strukturen, Hrsg.: H.-A. Richard, DVM e.V., ISSN 2509–8772 (2017) 127–136.

26. Siddique, S.; **Awd, M.;** Tenkamp, J.; Walther, F.: High and very high cycle fatigue failure mechanisms in selective laser melted aluminum alloys. Journal of Materials Research 32, 23 (2017) 4296-4304. https://doi.org/10.1557/jmr.2017.314

27. Siddique, S.; **Awd, M.;** Walther, F.: Influence of hybridization by selective laser melting on the very high cycle fatigue behavior of aluminum alloys. VHCF7, Proceedings of the 7th International Conference on Very High Cycle Fatigue, Dresden, ISBN 978–3–00–056960–9 (2017) 229–234.

28. Siddique, S.; **Awd, M.;** Tenkamp, J.; Walther, F.: Development of a stochastic approach for fatigue life prediction of AlSi12 alloy processed by selective laser melting. Engineering Failure Analysis 79 (2017) 34-50. https://doi.org/10.1016/j.engfailanal.2017.03.015

Related Presentations

1. **Awd, M.;** Kotzem, D.; Stern, F.; Teschke, M.; Tenkamp, J.; Walther, F. (V.): Fatigue damage tolerance assessment of materials and structures by enhanced testing and modeling approaches. AMC, European Assembly of Advanced Materials Congress, Stockholm, Sweden, 28.-31. Aug. (2022).

2. **Awd, M.;** Kotzem, D.; Stern, F.; Teschke, M.; Tenkamp, J.; Walther, F. (V.):Charakterisierung des Ermüdungsverhaltens und der Schädigungstoleranz additiv gefertigter Metalle.FGK-Online-Seminar, Forschungsinstitut für Glas-Keramik, Höhr-Grenzhausen, 20. June (2022).

3. **Awd, M. (V.);** Walther, F.: Reinforcement learning of VHCF fatigue strength in additively manufactured metals. AK Mikrostrukturmechanik / Working Group Microstructure Mechanics—11. May (2022).

4. **Awd, M.;** Kotzem, D.; Stern, F.; Teschke, M.; Tenkamp, J.; Walther, F. (V.): Ermüdungsverhalten und Schädigungstoleranz additiv gefertigter Metalle. 3. Fachtagung Werkstoffe und Additive Fertigung, Dresden und Online, 11.-13. Mai (2022).

5. Tenkamp, J.; **Awd, M.;** Stern, F.; Kotzem, D.; Walther, F. (V.): Coupled defect analysis and assessment of fatigue damage tolerance in additive manufactured metals. 5th Iberian Conference on Structural Integrity, Coimbra, Portugal, 30. Mar.-01. Apr. (2022).

6. **Awd, M. (V.);** Walther, F.: Influence of fatigue testing frequency on fracture and crack propagation rate in additive manufactured metals using the extended finite element method. 5th Iberian Conference on Structural Integrity, Coimbra, Portugal, 30. Mar.-01. Apr. (2022).

7. **Awd, M.;** Kotzem, D.; Stern, F.; Teschke, M.; Tenkamp, J.; Walther, F. (V.): Ermüdungsverhalten und Schädigungstoleranz additiv gefertigter Werkstoffe—Herausforderungen und Prüfstrategien. Werkstoffprüfung 2021, Web-Konferenz, 02.-03. Dez. (2021).

8. **Awd, M. (V.);** Walther, F.: Machine learning of fatigue strength of hybrid and additively manufactured aluminum alloys in VHCF regime. ASTM Symposium on Advances in Accelerated Testing and Predictive Methods in Creep, Fatigue, and Environmental Cracking. Virtual Event, 30. Nov.-2. Dec. (2021).

9. Tenkamp, J.; **Awd, M.;** Stern, F.; Kotzem, D.; Walther, F. (V.): Fatigue damage tolerance in additive manufactured metals applying enhanced measurement and modelling approaches. 46th MPA-Seminar 2021: Additive Manufacturing—Hydrogen—Energy—Integrity, Stuttgart, 12.-13. Oct. (2021).

10. **Awd, M. (V.);** Siddique, S.; Walther, F.: Microstructure-based assessment of damage mechanisms of selective laser melted Al-Si alloys under very high-cycle fatigue loading. VHCF8, 8th International Conference on Very High Cycle Fatigue, Online, Sapporo, Japan, 05.-09. July (2021).

11. **Awd, M. (V.);** Walther, F.; Siddique, S.; Fatemi, A.: Microstructure and fatigue damage evolution in additive-manufactured metals using enhanced measurement techniques and modeling approaches. TMS 2021 Annual Meeting & Exhibition, Online, USA, 15.-18. March (2021). **Invited Contribution.**

12. **Awd, M. (V.);** Stern, F.; Kotzem, D.; Tenkamp, J.; Walther, F.: Microstructure-based assessment and modeling of fatigue evolution and damage in additive manufactured metals. ASTM ICAM 2020, ASTM International Conference on Additive Manufacturing,

Online, 16.-20. Nov. (2020). **Keynote Lecture.**

13. **Awd, M. (V.);** Stern, F.; Kotzem, D.; Tenkamp, J.; Walther, F.: Multiscale experimental assessment and model-based correlation of fatigue damage evolution in additively man-ufactured metals. MSE 2020, Materials Science and Engineering, Darmstadt, 23.-25. Sept. (2020). **Keynote Lecture.**

14. **Awd, M. (V.);** Merghany, M.; Siddique, S.; Walther, F.: Simulation-based investigation of statistical fatigue strength of selective laser melted lightweight alloys. SIM-AM19, The Second International Conference on Simulation for Additive Manufacturing, Pavia, Italy, 11.-13. Sept. (2019). **Invited Contribution.**

15. **Awd, M. (V.);** Tenkamp, J.; Siddique, S.; Walther, F.: Additive manufacturing of metallic alloys—processes and properties. Kolloquium Zukunft Materialien, Hochschule Hamm-Lippstadt, Lippstadt, 23. May (2018). **Invited Contribution.**

16. Stern, F. (V.); Lüneburg, B.; Rollmann, J.; Pape, F.; Pantke, K. (V.); **Awd, M.;** Tenkamp, J.; Walther, F.: Evaluation of the influence of non-metallic inclusions on the high and very high cycle fatigue life of inductive hardened bearing steel in multi-megawatt wind turbines. Bearing World 2020, International Bearing Conference, Web Conference, 19.-23. Oct. (2020).

17. Tenkamp, J. (V.); **Awd, M.;** Siddique, S.; Walther, F.: Mechanism-based fatigue assess-ment and statistical property modeling of additively manufactured alloys in the HCF and VHCF regime. 4th Symposium on Structural Integrity of Additive Manufactured Materials & Parts, Oxon Hill, MD, USA, 07.-10. Oct. (2019). **Invited Contribution.**

18. Johannsen, J. (V.); **Awd, M.;** Siddique, S.; Emmelmann, C.; Walther, F.: Very High-Cycle Ermüdungseigenschaften und mikrostrukturelle Versagensmechanismen von laserstrahlgeschmolzenem (SLM) AlSi10Mg. Werkstoffwoche 2019, Dresden, 18.–20. Sept. (2019).

19. **Awd, M. (V.);** Merghany, M.; Johannsen, J.; Emmelmann, C.; Walther, F.: Microstruc-tural damage and fracture mechanisms of selective laser melted AlSi10Mg alloy under cyclic loading. ESIAM19, The First European Conference on Structural Integrity of Additively Manufactured Materials, Trondheim, Norway, 09.–11. Sept. (2019). **Invited Contribution.**

20. Siddique, S.; **Awd, M.;** Stern, F.; Tenkamp, J. (V.); Walther, F.: Effect of microstructure and internal defects on the cyclic deformation and damage behavior in additively (SLM) manufactured Al-Si alloys. TMS 2019 Annual Meeting & Exhibition, San Antonio, TX, USA, 10.–14. March (2019).

21. **Awd, M. (V.);** Siddique, S.; Johannsen, J.; Wiegold, T.; Klinge, S.; Emmelmann, C.; Walther, F.: Quality assurance of additively manufactured alloys for aerospace industry by non-destructive testing and numerical modeling. The 10th International Conference on Non-Destructive Testing in Aerospace, Dresden, Germany, 24.–26. Oct. (2018).

22. **Awd, M. (V.);** Siddique, S.; Hajavifard, R.; Walther, F.: Comparative study of defect-based and plastic damage-based approaches for fatigue lifetime calculation of selective laser-melted AlSi12. FFW 2018, 7th International Conference on Fracture Fatigue and Wear, Ghent, Belgium, 09.–10. July (2018).

23. **Awd, M. (V.);** Tenkamp, J.; Siddique, S.; Walther, F.: Additive manufacturing of lightweight metals – the process-property relationship. 10 Years ICAMS International Symposium, Bochum, 27.–25. June (2018).

24. **Awd, M. (V.)**; Johannsen, J.; Siddique, S.; Emmelmann, C.; Walther, F.: Qualification of selective laser-melted Al alloys against fatigue damage by means of measurement and modeling techniques. Fatigue 2018, 12th International Fatigue Congress, Poitiers, France, 27. May - 1. June (2018).

25. **Awd, M. (V.)**; Tenkamp, J.; Siddique, S.; Walther, F.: Additive manufacturing of metallic alloys—processes and properties. Kolloquium Zukunft Materialien, Hochschule Hamm-Lippstadt, Lippstadt, 23. May (2018).

26. Frömel, F. (V.); **Awd, M.**; Tenkamp, J.; Walther, F.; Boes, J.; Geenen, K.; Röttger, A.; Theisen, W.: Untersuchung des Einflusses einer HIP-Nachbehandlung auf den selektiv lasergeschmolzenen (SLM) austenitischen Stahl AISI 316L. Werkstoffe und Additive Fertigung, Potsdam, 25.-26. April (2018).

27. **Awd, M. (V.)**; Siddique, S.; Tenkamp, J.; Walther, F.: Freeform characterization of fatigue strength of additively manufactured lightweight alloys through FEM and Monte-Carlo modeling. 2. Tagung des DVM-AK Additiv gefertigte Bauteile und Strukturen, Berlin, 09.-10. Nov. (2017).

28. Siddique, S.; **Awd, M.**; Kampmann, A.; Tenkamp, J. (V.); Walther, F.: Defekt-basierte Charakterisierung des Ermüdungsverhaltens von laseradditiv gefertigten AlSi10Mg- und AlSi12-Legierungen. Werkstoffwoche 2017, Dresden, 27.-29. Sept. (2017).

29. Siddique, S. (V.); **Awd, M.**; Tenkamp, J.; Walther, F.: Very high cycle fatigue mechanisms and fatigue life prediction of additive manufactured Al-alloys. Hael Mughrabi Honorary Symposium & 28th Colloquium on Fatigue Mechanisms, Erlangen, 19.-21. July (2017).

30. Siddique, S. (V.); **Awd, M.**; Tenkamp, J.; Walther, F.: Influence of hybridization by selective laser melting on the very high cycle fatigue behavior of aluminum alloys. VHCF7, 7th International Conference on Very High Cycle Fatigue, Dresden, 3.-5. July (2017).

31. Tenkamp, J. (V.); Siddique, S.; **Awd, M.**; Kampmann, A.; Walther, F.: Experimental investigations on the effect of process-induced material characteristics on the mechanical performance of selective laser melted Al- and Ti-alloys. IKM-Seminar, Dresden, 10. Feb. (2017).

32. Siddique, S.; **Awd, M.**; Tenkamp, J. (V.); Walther, F.: Defect-correlated fatigue assessment of selective laser melted Al- and Ti-alloys in HCF- and VHCF-regions. International Symposium "Additive Manufacturing," Dresden, 08.-09. Feb. (2017).

33. **Awd, M. (V.)**; Siddique, S.; Imran, M.; Walther, F.: Development of a stochastic approach for fatigue life prediction of AlSi12 alloy processed by selective laser melting. MSE 2016, Materials Science and Engineering, Darmstadt, 27.-29. Sept. (2016).

Published Volumes

1. Tenkamp, J.: Charakterisierung und Modellierung des Ermüdungsverhaltens und der Schädigungstoleranz aushärtbarer Al-Si-Mg-Gusslegierungen im HCF- und VHCF-Bereich. Dissertation, Technische Universität Dortmund, Springer Vieweg Verlag, Wiesbaden (2022). https://doi.org/10.1007/978-3-658-38333-6

2. Scholz, R.: Mechanism-Based Assessment of Structural and Functional Behavior of Sustainable Cottonid. Dissertation, Technische Universität Dortmund, Springer Vieweg Verlag, Wiesbaden (2022). https://doi.org/10.1007/978-3-658-37540-9

3. Nebe, M.: In Situ Characterization Methodology for the Design and Analysis of Composite Pressure Vessels. Dissertation, Technische Universität Dortmund, Springer Vieweg Verlag, Wiesbaden (2022). https://doi.org/10.1007/978-3-658-35797-9

4. Hülsbusch, D.: Charakterisierung des temperaturabhängigen Ermüdungs- und Schädigungsverhaltens von glasfaserverstärktem Polyurethan und Epoxid im LCF- bis VHCF-Bereich. Dissertation, Technische Universität Dortmund, Springer Vieweg Verlag, Wiesbaden (2021). https://doi.org/10.1007/978-3-658-34643-0

5. Schmiedt-Kalenborn, A.: Mikrostrukturbasierte Charakterisierung des Ermüdungs- und Korrosionsermüdungsverhaltens von Lötverbindungen des Austenits X2CrNi18–9 mit Nickel- und Goldbasislot. Dissertation, Technische Universität Dortmund, Springer Vieweg Verlag, Wiesbaden (2020). https://doi.org/10.1007/978-3-658-30105-7

6. Wittke, P.: Charakterisierung spanlos gefertigter Innengewinde in Aluminium- und Magnesium-Leichtbauwerkstoffen. Dissertation, Technische Universität Dortmund, Springer Vieweg Verlag, Wiesbaden (2019). https://doi.org/10.1007/978-3-658-27943-1

7. Schmack, T.: Entwicklung einer ganzheitlichen Methode zur Bestimmung des dehnratenabhängigen Verhaltens faserverstärkter Kunststoffe. Dissertation, Technische Universität Dortmund, Springer Vieweg Verlag, Wiesbaden (2019). https://doi.org/10.1007/978-3-658-26931-9

8. Klein, M.: Mikrostrukturbasierte Bewertung des Korrosionsermüdungsverhaltens der Magnesiumlegierungen DieMag422 und AE42. Dissertation, Technische Universität Dortmund, Springer Vieweg Verlag, Wiesbaden (2019). https://doi.org/10.1007/978-3-658-25310-3

9. Siddique, S.: Reliability of Selective Laser Melted AlSi12 Alloy for Quasistatic and Fatigue Applications. Dissertation, Technische Universität Dortmund, Springer Vieweg Verlag, Wiesbaden (2019). https://doi.org/10.1007/978-3-658-23425-6

Printed in the United States
by Baker & Taylor Publisher Services